This series aims to report new developments in mathematical economics and operations research and teaching quickly, informally and at a high level. The type of material considered for publication includes:

1. Preliminary drafts of original papers and monographs
2. Lectures on a new field, or presenting a new angle on a classical field
3. Seminar work-outs
4. Reports of meetings

Texts which are out of print but still in demand may also be considered if they fall within these categories.

The timeliness of a manuscript is more important than its form, which may be unfinished or tentative. Thus, in some instances, proofs may be merely outlined and results presented which have been or will later be published elsewhere.

Publication of *Lecture Notes* is intended as a service to the international mathematical community, in that a commercial publisher, Springer-Verlag, can offer a wider distribution to documents which would otherwise have a restricted readership. Once published and copyrighted, they can be documented in the scientific literature.

Manuscripts

Manuscripts are reproduced by a photographic process; they must therefore be typed with extreme care. Symbols not on the typewriter should be inserted by hand in indelible black ink. Corrections to the typescript should be made by sticking the amended text over the old one, or by obliterating errors with white correcting fluid. Should the text, or any part of it, have to be retyped, the author will be reimbursed upon publication of the volume. Authors receive 75 free copies.

The typescript is reduced slightly in size during reproduction; best results will not be obtained unless the text on any one page is kept within the overall limit of 18 x 26.5 cm (7 x 10 ½ inches). The publishers will be pleased to supply on request special stationery with the typing area outlined.

Manuscripts in English, German or French should be sent to Prof. Dr. M. Beckmann, Department of Economics, Brown University, Providence, Rhode Island 02912/USA or Prof. Dr. H.P. Künzi, Institut für Operations Research und elektronische Datenverarbeitung der Universität Zürich, Sumatrastraße 30, 8006 Zürich.

Die „*Lecture Notes*" sollen rasch und informell, aber auf hohem Niveau, über neue Entwicklungen der mathematischen Ökonometrie und Unternehmensforschung berichten, wobei insbesondere auch Berichte und Darstellungen der für die praktische Anwendung interessanten Methoden erwünscht sind. Zur Veröffentlichung kommen:

1. Vorläufige Fassungen von Originalarbeiten und Monographien.
2. Spezielle Vorlesungen über ein neues Gebiet oder ein klassisches Gebiet in neuer Betrachtungsweise.
3. Seminarausarbeitungen.
4. Vorträge von Tagungen.

Ferner kommen auch ältere vergriffene spezielle Vorlesungen, Seminare und Berichte in Frage, wenn nach ihnen eine anhaltende Nachfrage besteht.

Die Beiträge dürfen im Interesse einer größeren Aktualität durchaus den Charakter des Unfertigen und Vorläufigen haben. Sie brauchen Beweise unter Umständen nur zu skizzieren und dürfen auch Ergebnisse enthalten, die in ähnlicher Form schon erschienen sind oder später erscheinen sollen.

Die Herausgabe der „*Lecture Notes*" Serie durch den Springer-Verlag stellt eine Dienstleistung an die mathematischen Institute dar, indem der Springer-Verlag für ausreichende Lagerhaltung sorgt und einen großen internationalen Kreis von Interessenten erfassen kann. Durch Anzeigen in Fachzeitschriften, Aufnahme in Kataloge und durch Anmeldung zum Copyright sowie durch die Versendung von Besprechungsexemplaren wird eine lückenlose Dokumentation in den wissenschaftlichen Bibliotheken ermöglicht.

Lecture Notes in Operations Research and Mathematical Systems

Economics, Computer Science, Information and Control

Edited by M. Beckmann, Providence and H. P. Künzi, Zürich

23

A. Ghosal

Department of Mathematics,
Monash University, Clayton/Australia

Some Aspects of Queueing and Storage Systems

Springer-Verlag
Berlin · Heidelberg · New York 1970

ISBN-13: 978-3-540-04947-0 e-ISBN-13: 978-3-642-88208-1

DOI: 10.1007/978-3-642-88208-1

Preface

In these lecture notes which grew out of my lectures at the Honours classes of Monash University, I have attempted to develop queueing and storage problems from a unified viewpoint. It has been recognized a decade ago that many of the queueing and storage problems belong to the same family of stochastic processes; so a problem in storage theory may have an analogue in a queueing situation and vice versa. In my notes I have highlighted this aspect and tried to develop a broad perspective in a student rather than to work out in detail the various exercises in queueing and inventory problems which are mostly available in the literature produced so far. I have also pointed out some of the practical applications of some theoretical results which appear to be important for an Operational Research worker. Many of the results given in these notes have cropped out of the author's own research over the last decade. Some new ideas which have scope for further exploitation have been given in most of the chapters (e.g. concept of cybernetic systems in Chapter 3, optimality problems in Chapter 4 and some problems in Chapter 2, etc.): these may benefit graduate students.

I thank my students for various discussions inside and outside the classrooms. I am also grateful to Mrs A. Darby for painfully and accurately typing out my manuscript.

Amitava Ghosal

September 1969

Index

CHAPTER 1

A UNIFIED TREATMENT OF QUEUEING AND STORAGE PROBLEMS - ELEMENTARY PROBABILITY DISTRIBUTIONS

1.1. A Simple Storage System

The level of a storage system (a reservoir, an inventory system, etc) depends on its input and its output (or release). In a simple storage model we have input from one source and only one form of output; we may have complex systems in which we have inputs of various types from various sources and the outputs are also of various forms going into various systems. We shall consider the simple system: if Z_t be the level at time t (which is for the time being taken to be discrete) before an input X_t goes into it and Y_t the release at the end of the time period (t,t+1), then we get

$$Z_{t+1} = [Z_t + X_t - Y_t]^+ \qquad (1.1)$$

suggesting that Z_t cannot be negative. If the storage system has a capacity k, we write

$$Z_{t+1} = \min[k, Z_t + x_t - Y_t]^+ \ . \qquad (1.2)$$

If either X_t or Y_t or both are random variables (r.v.) following some probability distribution functions (d.f.), then $\{Z_t\}$ is a random (or stochastic) process; the primary probability problem in the theory of storage is to develop methods of determination of the d.f. of $\{Z_t\}$ when the d.f.'s of X_t and Y_t are given. (See Moran (1954, 1955, 1958), Ghosal (1962a), Prabhu (1965)). It is also important to know for decision-making the probability that under a given situation, a storage system becomes empty (see Ghosal 1969,1964):

1.2. Analogy between Queueing and Storage Models.

In a queueing problem we are concerned with the d.f. of the waiting time of the customers and/or that of the queue size. Consider a single-server queueing model in which the interval between the arrival times of the nth and (n+1)th customers (inter-arrival time) is t_n, the service time of the nth customer is s_n and the waiting time of the nth customer (measured from his time of arrival to the time of the service starts) is w_n, then under a "first-come first-served" queue disciple we have the following relation binding w_{n+1} with w_n, t_n and s_n (see Lindley (1952))

$$w_{n+1} = [w_n + s_n - t_n]^+ . \tag{1.3}$$

If the customers decide not to wait beyond a time k (Finch (1960)), then we get

$$w_{n+1} = \min[k, w_n + s_n - t_n]^+ . \tag{1.4}$$

Again, if Q_t be the queue size at time t (excluding the one being served), A_t be the number of arrivals in the interval (t,t+1), B_t the number served during the same period, then Q_{t+1} the queue size at tme t + 1 is given by

$$Q_{t+1} = [Q_t + A_t - B_t]^+ . \tag{1.5}$$

From the nature of equations (1.1), (1.3) and (1.5) we find that basic stochastic model for all of these is the same. We express both (1.3) and (1.5) by (1.1) if we define Z_t, X_t and Y_t according to Table 1.1 given below:

TABLE 1.1 A Unified Model

$$Z_{t+1} = [Z_t + X_t - Y_t]^+$$

Notation	Storage System (Dam, Inventory, etc.)	Queues: Waiting Time Model	Queues: Queue Size Model
Z_t	Storage level at the beginning of the interval (t,t+1)	Waiting time of the t-th customer	Queue Size at arbitraty time t
X_t	Input at the beginning of the interval	Service time of the t-th customer	Arrivals during interval (t,t+1)
Y_t	Release at the end of the interval	Inter-arrival time between t-th and (t+1)th customers	Departures after service during interval (t,t+1)

With the help of Table 1.1 we can interpret results in storage model in terms of queueing models and vice-versa.

For a queue-size model in a single-server queue, we can define Z_t as the queue size left by the t-th departing customer, X_t arrivals during inter-departure between t-th and (t+1)th departures, then $Y_t = 1$, so that $Z_{t+1} = [Z_t + X_t - 1]^+$.

We can also represent a bulk-service queue model (Bailey (1954)) by the model (1.1) Here the service is rendered to a group of Y_t customers at a time (usually

$Y_t = m$, a constant); Z_t would represent the queue size after the departure of the t th group of customers and X_t the number of arrivals during the inter-departure interval between the t th and (t+1)th group.

It will be seen thus that most of the queueing and storage models can be given a structure as that given in Table 1.1. There are however a few situations in which we do not get exactly the analogue of a model (developed in a queueing case) in a storage case, and vice versa. For example, queues in series do not have an analogue for dams in series (this topic will be discussed later on in the notes).

1.3 Discrete and Continuous Cases

In the treatment given so far we have considered time (t) to be descrete (t = 0,1,2,...). For discrete time, Z_t can have a continuous d.f. if both X_t and Y_t follow continuous d.f.'s; similarly if X_t and Y_t both follow discrete d.f.'s, Z_t will also have a discrete d.f. (subject to the condition that the d.f. of Z_t exists). For example, when we are dealing with the storage of items like cars, aircraft parts, etc., Z_t, X_t and Y_t have all discrete d.f.'s while in the case of dams Z_t will have a continuous d.f. Similarly in the waiting time model (1.3) we deal with continuous d.f. of $\{w_n\}$, while in (1.5) the queue size Q_t has a discrete d.f. When Z_t has a continuous d.f. we denote its d.f. as

$$F_t(y) = \text{prob}(Z_t \leq y), \qquad 0 \leq y \leq k, \tag{1.6}$$

if the range of variation of y is [0,k] - k can be finite or infinite. When Z_t has a discrete d.f. we denote it as

$$F_i^{(t)} = \text{Prob}(Z_t = i), \qquad i = 0,1,\ldots,k, \tag{1.7}$$

k can be finite or infinite. If Z_t attains a limiting d.f. which is independent of t we get

$$\lim_{t\to\infty} F_t(y) \to F(y) \tag{1.8}$$

$$\lim_{t\to\infty} F_i(t) \to F_i \ . \tag{1.9}$$

We call (1.8) and (1.9) as limiting or stationary d.f. of Z_t.

We can also treat time (t) as continuous; a continuous t is very relevant in a production system or in a dam. In queueing models we are mostly concerned with discrete time. The continuous analogue for the model (1.2) is given by

$$Z(t+\delta t) = \min[k, Z(t) + \delta X(t) - \delta Y(t)]^+ \tag{1.10}$$

where Z(t) is the level of the storage system at time t, $\delta X(t)$ the increment to stock during time interval (t, t+δt) and $\delta Y(t)$ the amount released (or consumed) during the same infinitesimal period. When $k = \infty$, we get the continuous analogue for (1.1). Work on continuous time storage models has been done by various authors (for a detailed reference see Prabhu (1965)).

1.4 Kendall's classification

David G. Kendall (1953) gave a classification for queueing problems with the following definitions:

(i) GI/G/s GI implies that the input X_t (in Table 1.1) follows a general d.f G(x);

G implies that the output Y_t (in Table 1.1) follows a general d.f. H(x);

s implies that the number of servers is s

(s = 1,2,...).

(ii) M/G/s The same as GI/G/s with the exception that the input X_t follows a Markovian d.f. (exponential d.f. for the inter-arrival of customers in a queue).

(iii) GI/M/s The same as GI/G/s with the exception that the output Y_t follows a Markovian d.f. (in the sense of (ii)).

From the above it is easy to decode the systems e.g. M/M/1, GI/M/1, M/G/1, etc. To specify the d.f. of X_t further, he defined E_p/G/s to be a system for which the inter-arrival time follows an Erlangian d.f. given by

$$dG(t) = \frac{(\lambda p)^p}{(\lambda-1)!} e^{-\lambda pt} t^{p-1} dt \quad (0 \leq t < \infty \ \lambda > 0). \tag{1.11}$$

When p = 1, we get the exponential d.f. Again when p = ∞, the inter-arrival time is constant (1/λ). If the input is constant we denote the system as D/•/•; if the output is constant we denote it by •/D/• .

1.5 Principal problems

Having set the queueing and storage models in the form of a unified model (Table 1.1), our principal problems are as follows:

(i) given the d.f.'s of X_t and Y_t, to determine the d.f. of Z_t;

(ii) given the d.f.'s of X_t and Y_t and a given $Z_o = y$ (say), to find the probability that Z_t subsequencly becomes empty before reaching the capacity level, and also to find the time over which the system is not empty (this is the busy period phase in a queueing model);

(iii) given cost factors governing a queueing or a storage model and the d.f.'s of X_t and Y_t, to determine the optimal capacity (k) of the system to satisfy some criteria, e.g. minimum cost of maintenance, maximum net proceeds from the sales over a time horizon.

From the point of view of determination of the d.f. of Z_t, we can denote a system represented by the model (1.2) as $\langle Z_t, X_t, Y_t, k \rangle$. It can be realised easily from the linear model (1.1) or (1.2) that the d.f. of Z_t depends on the d.f. of $u_t = X_t - Y_t$ rather than on that of the individual d.f.'s of X_t and Y_t; so it is sufficient to represent the system (1.2) as $\langle Z_t, u_t, k \rangle$.

1.6 Existence of the limiting d.f. of Z_t

In a system $\langle Z_t, u_t, k \rangle$, the problem (i) referred to in Section 1.5 implies that given the d.f. of u_t we are to determine the d.f. of Z_t. In many of the problems it is of interest to investigate whether the limiting d.f. F(y) of Z_t exists (see eqn (1.8)). Lindley (1952) has proved that, if $k = \infty$, $F(y) = \lim_{t\to\infty} F_t(y)$ of Z_t exists if the expectation $E(u_t) < 0$. The physical interpretation of this result can be easily understood if we write (1.1) for $k = \infty$ in the following form

$$Z_{t+1} = [Z_t + u_t]^+ \tag{1.12}$$

where $-\infty < u_t < \infty$. If Z_t has not been zero over the time interval (0,t), then given $Z_o = 0$,

$$Z_t = \sum_{i=1}^{t} u_i ;$$

we can therefore write (1.12) in the form

$$Z_{t+1} = Z_r + \sum_{i=r}^{t} u_i ,$$

so that

$$Z_t = \sup_{0<m\leq t} \sum_{i=1}^{m} u_{t-i} \tag{1.13}$$

Lindley has proved that Z_t converges to an infinite r.v. $Z = \sup_{m>o} \sum_{i=1}^{m} u_i$ (as t tends to infinity) only if $E(u_i) < 0$ (see also Section 1.7 below). From physical reasoning we find that if $E(u_i) \geq 0$, then $\sup_{m>o} \sum_{i=1}^{m} u_i$ diverges and tends to infinity; in a queueing situation this would mean that the queue length would be infinite in the long run. When k is finite, we have an automatic control process given by

$$Z_{t+1} = \min[Z_t + u_t, k]^+ \tag{1.14}$$

implying that even if $Z_t + u_t$ exceeds k, Z_{t+1} is kept at the level k. For such a system the limiting d.f. of Z_t exists if $E(u_t) < \infty$ (see Kendall (1951), Finch (1958), Finch (1960)).

While working on the basis of (1.13) our basic assumption that the d.f. of $U_{r,m} = \sum_{i=r+1}^{r+m} u_i$ is the same for any r. The limiting d.f. of Z_t, both for finite and infinite k, is the same as that of the cumulative sum $Z = \sup_{m>o} U_m$, if such a limit exists. This holds true both for discrete and coninuous Z_t.

For continuous time we write Z(t) for Z_t, u(t),for u_t U(t) for U_t, and get the same result. The limiting d.f. of Z(t) is the same as that of $Z = \sup_{m>o} U(m)$, if the limiting d.f. exists. For continuous t, U(t) is a homogeneous separable stochastic process, implying that $U(r,t) = \int_r^{r+t} u(m)dm$ is independent of r, and secondly that if $t = t_1 + t_2$, $prob(U(t) \leq x) = prob(U(t_1) + U(t_2) \leq x)$.

In a queueing system the waiting time process $\{w_t\}$ and the queue size prooess $\{Q_t\}$ constitute a combined process $\{Z_t\} = \{w_t, Q_t\}$; so if either w_t or Q_t is bounded, the d.f.'s of both w_t and Q_t exist. This aspect will be discussed in greater detail subseqeuntly (see Ghosal (1968). Section 1.11 of these notes, etc.).

1.7 Spitzer's Identity

In the foregoing section we have found that the d.f. of Z_t depends primarily on the d.f. of U_m; the same holds true when the time t is continuous. A logical approach is therefore to find a suitable way of relating the d.f. of Z_t to that of U_m. A general solution of this problem is not available so far; but for the case $k = \infty$, and for certain conditions the basic relationship is provided by the identity discovered by Spitzer (1956). Spitzer's identity provides a relationship between the characteristic function (c.f.) of $\sup_{o \leq m \leq t} U_m$ and the c.f. of $U_t^+ (= \max(0, U_t))$, provided u_t's are mutually independent and follow the same d.f. The identity can be expressed as below (for $|\gamma| < 1$):

$$\sum_{t=o}^{\infty} \phi_t(\theta)\gamma^t = \exp\left[\sum_{t=1}^{\infty} \frac{\psi_t{+}(\theta)}{t} \gamma^t\right] \qquad (1.15)$$

where $\phi_t(\theta)$ = c.f. of $\max[0, U_1, \ldots, U_t]$

$$= \int_{W_t} e^{i\theta W_t} dF(W_t)$$

$$(W_t = \max(0, U_1, \ldots, U_t), \quad U_o = 0,)$$

and $\psi_t^+(\theta)$ = c.f. of U_t^+

$$= \int_{U_t} e^{i\theta U_t^+} dG(U_t^+),$$

$F(\cdot)$ and $G(\cdot)$ being the d.f. of W_t and U_t^+ respectively. Spitzer's result (1.15) generalizes the result of Sparre-Andersen (1954), Theorem 1). Spitzer proved (1.15) by resorting to combinatorial methods; later on Wendel (1958) gave a shorter proof through the aid of Banach algebra.

It was first noticed by Pollaczek (1957) and also proved by Sptizer (1958) that the d.f. of the variable Z_t given by (1.12), in the limiting case, is the same as that of $W_t = \max(0, U_1, \ldots, U_t)$. Putting in the symbolic form we write

$$Z_t \sim W_t = \max(0, U_1, \ldots, U_t), \qquad (1.16)$$

where "$\sim$" stands for "equivalence in d.f." To prove it, let us put recursively

$$U_1' = u_t^+ \,, \qquad\qquad = \max(0,u_t)$$

$$U_2' = (U_1' + u_{t-1})^+, \qquad\qquad = \max(0,u_t,u_{t-1})$$

$$U_3' = (u_{t-2}+ (U_1' + u_{t-1})^+)^+,$$

$$U_t' = (u_1+ (u_2+\ldots+(U_1' + u_{t-1})^+)^+\ldots)^+ \quad = \max(0,u_t,u_{t-1},\ldots,u_1). \qquad (1.17)$$

Then we have, since $u_{t-r+1} \sim u_r (r = 1,\ldots,t)$, and $(U_{t-1}' + u_1)^+$,

$$U_t' \sim W_t \qquad (1.18)$$

It remains to show that, since U_t' follows the same model as that of Z_t, $U_t' \sim Z_t$. Spitzer (1958) proves it as follows. If we assume

$$\sum_{k=1}^{\infty} \frac{1}{k} \Pr(U_k > 0) < 0$$

then

$$\lim_{t\to\infty} \max_{o\leq k\leq t} U_k = \max_{k\geq o} U_k < \infty \qquad (1.19)$$

implies that the sequence W_t as defined above converges to an r.v. (on Borel space Ω) with probability one (in Ω). Hence W_t converges to W in distribution. Similarly Z_t as defined in (1.12) also converges to Z in distribution; since both W_t and Z_t converge to a single point, $W \sim Z$. Thus we have proved that

$$Z_t \sim U_t' \,; \qquad (1.20)$$

hence in (1.15) $F(\cdot)$ must be the d.f. of Z_t.

Differentiating both sides of (1.15) w.r.t. θ and setting $\theta = 0$, we get

$$E\{\max_{o\leq m\leq t} U_m^+\} = \sum_{m=1}^{t} \frac{1}{m} E(U_m^+). \qquad (1.21)$$

Again, by letting $\theta \to \infty$ in (1.15) we get

$$\sum \Pr\{\bigcap_{m=1}^{t} (U_m \geq 0)\}\gamma^t = \exp\{\sum_{m=1}^{\infty} \frac{\gamma^m}{m} \Pr(U_m \geq 0)\}; \qquad (1.22)$$

where $\bigcap_{m=1}^{t} (U_m > 0) \Rightarrow$ the event $\{Z_t > 0\}$.

It may be remarked that the relationship (1.15) does not directly relate the p.d.f. of $\max_m U_m$ to that of U_m. We can however derive a relation between $\phi_t(\theta)$ and $\psi_t(\theta) = E(e^{i\theta U_t})$, the c.f. of U_t irrespective of sign. In Spitzer (1957, eqn. (3.3) p. 332) we find that

$$\psi_t^+(\theta) - 1 = \lim_{\varepsilon \to o} \int_{-\infty}^{\infty} \delta_\varepsilon(x) d\,Pr(U_t \leqslant x) \tag{1.23}$$

where $\delta_\varepsilon(x) = e^{i\theta x} - e^{i\varepsilon x}$ for $x \geqslant 0$, $\delta_\varepsilon(x) = 0$ for $x < 0$, $\theta > 0$; $\varepsilon > 0$.

Hence we get

$$\psi_t^+(\theta) - 1 \to \int_o^\infty e^{i\theta x} dPr(U_t \leqslant x) + Pr(U_t \leqslant 0),$$

whence we get from (1.15)

$$(1-\gamma) \sum_{t=o}^{\infty} \phi_t(\theta)\gamma^t \to \exp[\sum_{t=1}^{\infty} \frac{\gamma^t}{t} \{\psi_t(\theta) + Pr(U_t \leqslant 0\}]. \tag{1.24}$$

If Z_t has a limiting d.f. and $\lim_{t\to\infty} \phi_t(\theta) \to \phi(\theta)$, then we get,

$$\phi(\theta) = \exp[\sum_{t=1}^{\infty} \frac{\gamma^t}{t} \{\psi_t(\theta) + Pr(U_t < 0)\}] ; \tag{1.25}$$

the term on R.H.S. will have to be free of γ for (1.25) to hold good. In a few queueing problems (say M/M/1) it is easier to get the d.f. of waiting time by applying (1.25) rather than by (1.15). It is also easily seen that when γ tends to 1, $\phi_t(\theta)$ tends to $\phi(\theta)$, and we get the expression (1.25) in which $\gamma = 1$. It can be verified that (1.25) holds true only if γ tends 1.

The other way of getting round the problem of determining $\psi_t^+(\theta)$ in (1.15) is to express it in terms of $\psi_t(\theta)$ in Pollaczek's way (eqn.(1.14) in Pollaczek(1957) who used Laplace transform instead of Fourier transform used by us here - but the result remains true):

$$E(e^{-\theta U^+}) = \frac{1}{2\pi i} \int_C E(e^{-\zeta U}) \frac{\theta}{\zeta(\theta-\zeta)} d\zeta \qquad (1.26)$$

where $0 < \mathrm{Re}(\zeta) < \mathrm{Re}(\theta)$, and the contour is a line parallel to the imaginary axis at a distance $\mathrm{Re}(\zeta)$ and within the analytical region of $E(e^{-\zeta U})$ (see also Le Gall (1962), p.32).

Spitzer's identity has a great theoretical interest in the study of queueing and storage models of GI/G/1 type (for further reference, see Kingman (1962, 1966), Finch (1961)). It is however difficult to get exact d.f. of waiting time or storage level by applying this identity. Baxter and Donsker (1958) have developed a continuous analogue for Spitzer's identity by applying it on a separable stochastic process {Z(t) and U(t)}. There is scope for extending this identity to cover finite processes, viz. $Z_t < k$ for the model (1.14). Intuitively, it appears that if we define $\hat{U}_t$ as follows

$$\hat{U}_t = \min[k, U_t^+], \qquad (1.27)$$

then (1.15) would reduce to

$$\sum_{t=o}^{\infty} \gamma^t E(e^{i\theta Z_t}) = \exp[\sum_{t=1}^{\infty} \frac{\gamma^t}{t} E(e^{i\theta \hat{U}_t})]$$

and in the limiting case $(Z_t \to Z)$,

$$E(e^{i\theta Z}) = \lim_{t\to\infty} \exp[\sum_{t=1}^{\infty} \frac{1}{t} E(e^{i\theta \hat{U}_t} - 1)] \quad (0 \leq z \leq k) \qquad (1.28)$$

The result (1.28) is a conjecture. The R.H.S. of (1.28) will have to be in a form so that it is independent of γ; it remains true iff $\gamma = 1$.

1.8 Wiener-Hopf Decomposition

If for the model (1.12), the d.f. of u_t be $G(x)$ $\{= Pr(u_t \leq x)$ for all $t\}$, u_t being continuous, and the d.f. of Z_t be $F_t(y)$ $\{= Pr(Z_t \leq y)\}$, then we get the following integral equation (Z_t and u_t being independent, and u_t's being mutually independent)

$$F_{t+1}(y) = -\int_0^{\infty} F_t(x)dG(y-x) \qquad (0 \leq y < \infty) \qquad (1.29)$$

so that where $\lim_{t\to\infty} F_t(y) = F(y)$, we get

$$F(y) = -\int_0^{\infty} F(x)dG(y-x) \qquad (0 \leq y < \infty). \qquad (1.30)$$

It may be remarked that the d.f. $F(\cdot)$ has a concentration at zero. The eqn (1.30) for determining $F(\cdot)$ from the knowledge of $G(\cdot)$ is a Wiener Hopf integral equation, and was first used by Lindley (1952) is deriving the d.f. of waiting time in a GI/G/1 queue. Apparently Lindley's technique can be applied for studying storage models in which Z_t is unbounded on right. The condition for the existence of a unique solution of $F(y)$ in (1.30) is that $E(u) = \int_{-\infty}^{\infty} xdG(x) < 0$ (see Lindley (1952)); it has also been found by Spitzer (1957, p.342) that the variance of u, $\sigma^2(u) = \int_{-\infty}^{\infty} \{x-E(x)\}^2 dG(x)$ should be finite. The interpretation of (1.30) (also of model (1.12)) is that

$$Z \sim [Z + u]^+ . \qquad (1.31)$$

We shall elucidate Lindley's method in applying in the queueing system M/G/1 and then on $M/E_n/1$. An interested reader can compare the merits of Lindley's method and Spitzer's method in deriving the waiting time d.f. of queueing systems. It may however be remarked that Lindley's method does not yield practical solution to all GI/G/1 systems (see also Smith (1953)).

Lindley derives the relationship between the c.f. of Z_t (waiting time of the t th customer or the storage level of an infinite dam at time t) and that of $u_t(= X_t - Y_t)$ in the following manner. Let us define

$$\phi(\theta) = \int_{-0}^{\infty} e^{i\theta y} dF(y), \quad \psi(\theta = \int_{-\infty}^{\infty} e^{i\theta y} dG(y) \tag{1.32}$$

where $F(y) = \Pr\{Z_t \leqslant y\}$, $G(y) = \Pr\{u_t \leqslant y\}$. Let us also define a function $F^*(y)$ in the interval $(-\infty,\infty)$ such that in $(0,\infty)$ $F^*(y) = F(y)$; in the range $(-\infty,0)$ $F^*(y)$ represents the d.f. of the idle time of the server. Hence

$$\int_{-\infty}^{\infty} e^{i\theta y} dF^*(y) = \int_{-\infty}^{0} e^{i\theta y} dF^*(y) - F(0) + \int_{-0}^{\infty} e^{i\theta y} dF(y)$$

$$= \int_{-\infty}^{0} e^{i\theta y} dF^*(y) - F(0) + \phi(\theta), \tag{1.33}$$

where $F(0) = \Pr\{Z_t \leqslant 0\}$. From the equation (1.30) we find that the L.H.S. (left hand side) of (1.33) is equivalent to $\phi(\theta)\psi(\theta)$, so that from (1.33) we get

$$\phi(\theta)\psi(\theta) = \int_{-\infty}^{0} e^{i\theta y} dF^*(y) + F(0) + \phi(\theta)$$

or $$\phi(\theta)\{\psi(\theta) - 1\} = \int_{-\infty}^{0} e^{i\theta y} dF^*(y) - F(0). \tag{1.34}$$

The equation (1.34) forms the basis of determining F(y) from the knowledge of G(y). The cases in which we are able to solve (1.34) are mostly the systems of type M/G/1. We may compare (1.34) with the Spitzer's identity (1.15) or (1.25).

1.9 D.F. of M/G/1 Systems

We shall apply Lindley's method of integral equations to solve for the d.f. F(y) of Z_t in the limiting case. Let the arrival time (Y_t) d.f. be $\lambda \exp(-\lambda x)dx$ $(0 \leqslant x < \infty;\ \lambda > 0)$, $G_1(x)$ the d.f. of service time (X_t), then the then the d.f. $G(x)$ of $u_t (= X_t - Y_t)$ is given by

$$G(x) = \int_0^\infty e^{-\lambda z} G_1(x+z)dz.$$

Then from the definition of $F^*(\cdot)$ given in Section 1.8, we find that for all $y < 0$ (from (1.3))

$$F^*(y) = \lambda \int_{x \leq y} \int_{z \geq 0} e^{-\lambda z} F(y-x) dG_1)(x+z)dz$$

$$= \lambda \int_{w \geq 0} \int_{v \geq 0} e^{-\lambda(w+v-y)} F(w) dG_1(v) dw$$

$$= C\, e^{\lambda y} \tag{1.35}$$

where C is constant (see Lindley (1952),p.283). From the equations (1.34) and (1.35), we get, if $\zeta(\theta) = E(e^{i\theta X_t})$, the c.f. of X_t,

$$\phi(\theta) = C[1 + \frac{1-\zeta(\theta)}{i\theta}]^{-1} \tag{1.36}$$

We determine c from the condition that $\phi(0) = 1$; hence because $E(Y) = 1/\lambda$ and

$$\lim_{\theta \to 0} \frac{1-\zeta(\theta)}{i\theta} = -E(X),$$

it follows that

$$C = 1 - \frac{E(X)}{E(Y)} = 1 - \rho \tag{1.37}$$

where $\rho = E(X)/E(Y)$. Thus $C = F(0)$.

Lindley's method developed above is particularly suited for application to $M/E_k/1$ systems in which the service time (X_t) follows Erlangian (or Gamma) distribution. When the service time is of more general type, we derive $\phi(\theta)$, the c.f. of Z_t, from (1.34), and compute mean, variance and if necessary moments of higher order for the d.f. of Z_t.

Examples

(i) M/M/1. Let the d.f. of X_t (service time or dam input) be exponential,

$$\Pr(X_t \leq x) = 1 - \mu e^{-\mu x} \quad (0 \leq x < \infty;\ \mu > 0) \tag{1.38}$$

The condition for the existance of a limiting d.f. of Z_t is that $E(X) < E(Y)$ or $\lambda < \mu$. The d.f. $G(x)$ of u_t is given by

$$G(x) = \begin{cases} 1 - \frac{\lambda}{\lambda+\mu} \exp(-\mu x) & (x>0), \\ \frac{\mu}{\lambda+\mu} \exp(\lambda x) & (x\leq 0) \end{cases} \tag{1.39}$$

so that

$$\psi(\theta) = \lambda\mu/(\mu-i\theta)(\lambda+i\theta). \tag{1.40}$$

From (1.37) we find $C = 1 - (\lambda/\mu)$, so that from (1.34) and (1.33) (we can also obtain $\phi(\theta)$ directly from (1.34))

$$\phi(\theta) = (1 - \frac{\lambda}{\mu}) \frac{\mu-i\theta}{\mu-\lambda-i\theta}$$

$$= (1 - \frac{\lambda}{\mu}) + \frac{\lambda(\mu-\lambda)}{\mu(\mu-\lambda-i\theta)} \quad , \tag{1.41}$$

whence we get

$$F(y) = \begin{cases} 1 - \lambda/\mu & (y=0), \\ 1 - \lambda/\mu \quad \exp -(\mu-\lambda)y\} & (0 < y < \infty) \end{cases} \tag{1.42}$$

(ii) $M/E_p/1$.

Let the d.f. of the service time (X_t) be

$$dG_1(x) = \frac{\mu^p}{(p-1)!} e^{-\mu x} x^{p-1} dx \quad (\mu > 0;\ 0 \leq x < \infty). \tag{1.43}$$

We derive the d.f. $G(x)$ of u_t (= $X_t - Y_t$) as follows:

$$G(x) = \Pr\{u_t \leqslant x\} = \int_z \Pr(X_t < x + Y_t) \Pr\{z \leqslant Y_t \leqslant z + dz\}$$

$$= \begin{cases} \lambda\int_0^{\infty}\{1 - e^{-\mu(x+z)}\sum_{r=0}^{p-1} \frac{\mu^r(x+z)^2}{r}\}e^{-\lambda z}\, dz & (x > 0) \\ \lambda\int_{-x}^{\infty}\{1 - e^{-\mu(x+z)} \sum_{r=0}^{p-1}\frac{\mu^r(x+z)^r}{r!}\}e^{-\lambda z} & (x \leqslant 0); \end{cases}$$

whence we get

$$G(x) = \begin{cases} 1 - e^{-\mu x}\{1 - (\mu/\lambda+\mu)^p\}\sum_{r=0}^{p-1}(\lambda+\mu)^{r+1}x^r/r! & (x > 0), \\ e^{\lambda x}\ 1 - (\mu/\lambda+\mu)^p & (x \leqslant 0) \end{cases} \qquad (1.44)$$

We can now derive $\psi(\theta)$ from (1.4) from the following

$$\psi(\theta) = \int_0^{\infty} e^{i\theta x} dG(x) + \int_{-\infty}^{0} d^{i\theta x} dG(x),$$

and then make use of (1.34) and (1.35) to obtain $\phi(\theta)$. Alternatively we can derive $\phi(\theta)$ directly from (1.36) and (1.37), viz.

$$\phi(\theta) = 1 - \frac{p\lambda}{\mu}\ [1 + \lambda\{1 - \mu^p/(\mu - i\theta)^p\}/i\theta]^{-1} \qquad (1.45)$$

In any M/G/1 system, we get from (1.32)

$$\phi(\theta)\{\psi(\theta) - 1\} = F(0)\lambda/(\lambda + i\theta) \qquad (1.46)$$

so that if we know $\psi(\theta)$ we can determine $\phi(\theta)$. The problem ultimately is to invert $\phi(\theta)$ to obtain $F(y)$.

[N.B. Lindley (1952, pp 286-7) has shown that for systems of $E_p/G/1$ type, we get the solution of $F(y)$ in the following form

$$F(x) = 1 + \sum_{i=1}^{N} c_i \exp(z_i x) \qquad (1.47)$$

where the c_i are complex constants. It may however be mentioned here that asymptotically any solution to the Wiener-Hopf equation behaves like the exponential function at infinity.]

(iii) Idle Time of the Server in M/G/1

The idle time of the server in an M/G/1 system corresponds to the situation when the waiting time Z_{t+1} as given by the relation $Z_{t+1} = Z_t + u_t$ is negative, but since it cannot be negative it is zero. Thus if we define I_t is the idle time of the server after serving the t th customer, we get from the model (1.12) (or (1.1) or (1.3)).

$$I_t = - Z_{t+1} = - [0, Z_t + u_t]^- . \qquad (1.48)$$

The interpretation of I_t in the storage model is that it represents shortage at the beginning of the time interval (t+1,t+2) (see Table 1.1).

From Lindley's result (1.33), if the inter-arrival time Y_t follows the exponential d.f. with parameter λ, then, for any d.f. of the service time X_t

$$F^*(y) = \Pr\{I_t \geq y\}$$

$$= C\, e^{-\lambda y} \quad (\lambda > 0, \quad y < \infty, \ c > 0) \qquad (1.49)$$

where $C = 1 - E(X)/E(Y) = 1 - \lambda E(X)$. The idle time d.f., therefore, is exponential with the same parameter as that of the inter-arrival time with a concentration at zero given by

$$\Pr(I_t = 0) = \Pr(Z_t > 0) = 1 - C = 1 - \lambda E(X). \qquad (1.50)$$

In terms of storage theory, (1.49) implies that the shortage follows an exponential distribution, if the a release (from an inventory or a dam) follows an exponential distribution, no matter what the d.f. of the input is, with a concentration at zero given by (1.47). This result is an extremely useful result in a control process which can be schematically demonstrated in Fig. 1.1.

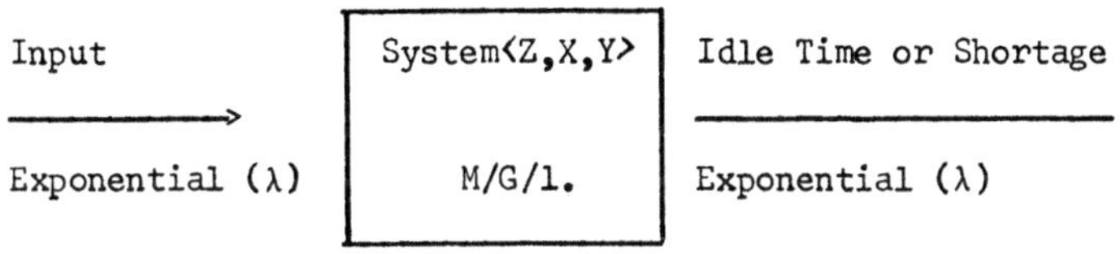

Figure 1.1. Shortage as output in the Control Process.

The equation (1.49) can be utilised in calculating the expeceed penalty cost resulting from shortage in an inventory model of M/G/1 type; when back-logging is allowed in an inventory situation, (1.49) enables us to estimate the expected amount of back-logged demand given by

$$E(I) = C/\lambda^2$$
$$= \{1 - \lambda E(X)\}/\lambda^2. \qquad (1.51)$$

Similarly we can calculate the variance of I, which is of practical significance in many operational research (O.R.) problems.

(iv) M/D/1 System

In an M/G/1 system if the service time (input in storage system) is constant m (say), we get

$$\Pr(u_t \leqslant x) = \Pr(Y_t \geqslant m - x)$$
$$= e^{-\lambda(m-x)} \quad (0 \leqslant x \leqslant m), \quad (= 1 \text{ for } x > m)$$

so that

$$\psi(\theta) = E(e^{i\theta x})$$

$$= \{\lambda \exp(-\lambda m)/(\lambda+i\theta)\}\{1- \exp(\lambda+\theta i)m\},$$

so that we can make use of (1.46) in which $F(0) = 1 - \lambda m$. Hence we find that Z_t follows an exponential distribution with a concentration at zero.

Alternatively we can obtain $\phi_\alpha(\theta)$ the L.S. transform of Z_t from $\zeta_\alpha(\theta)$, the L.S. transform of Y_t by using (1.37), then apply Lindley's method (1952,pp.286-7) by putting (see also (1.47))

$$F(y) = \sum_i c_i \exp(-z_i y)$$

where c_i are complex constants. This method was used by Gani and Prabhu (1957) for deriving the d.f. of Z_t in a D/M/1 system. The practical application of M/D/1 model is in inventory situations for which the demand is exponential and the supply is deterministic.

1.10 D.F. of Z_t in Finite Systems

In Section 1.8 and 1.9 we discussed the method of integral equations to derive the limiting d.f. of Z_t in model (1.1) where $k = \infty$. In this section we shall discuss the method of deriving F(y), the d.f. of Z_t when k is finite (model (1.2)) and Z_t and u_t are continuous r.v.'s. When k is finite (in other words the capacity of a dam or storage system is finite), then (1.30) will reduce to

$$F(y) = G(y-k) - \int_0^k F(x)dG(y-x), \quad (0 \leq y \leq k) \tag{1.52}$$

where $G(\cdot)$ is the d.f. of $u_t(= X_t - Y_t)$ and $F(y) = 1$ for $y \geq k$. This model corresponds to a random walk with two impenetrable barriers at 0 and k; the integral equation (1.52) is of Fredholm's type, and a unique solution exists if (i) G(x) is absolutely continuous and G(y-k) is non-zero fro some x in (1.49), (ii) $\int_{-k}^{k}(k - |y|)g^2(y)dy < 0$ where g(•) is the derivative of G(•). There is no solution to (1.52) if the term G(y-k) on the R.H.S. of (1.52) does not occur (see Ghosal (1964)). It can be seen that the integral equation (1.52) is the continuous analogue of a system of simultaneous equations.

It may be worthwhile investigating whether Lindley's method (equations (1.33) and (1.34) developed for $k = \infty$ in Section 1.8, can be extended to the class of problems represented by (1.52). A possible approach would be to define $F^*(y)$ in the interval $(-\infty,\infty)$ such that in $(0,k)$ $F^*(y) = F(y)$; in the range $(-\infty,0)F^*(y)$ would represent the shortage (or idle time) d.f., $F(0) = \int_{-\infty}^{0} dF^*(y)$; in the range (k,∞) $F^*(y-k)$ represents the d.f. of spillage (overflow) with $F(k) = \int_{k}^{\infty} dF^*(y)$.

Possibly the analogue of (1.34) would be

$$\phi(\theta)\{\psi(\theta) - 1\} = \int_{-\infty}^{0} e^{i\theta y} dF^*(y) + \int_{k}^{\infty} e^{i\theta y} dF^*(y) - F(0) - e^{i\theta k} f(k). \qquad (1.53)$$

In this case the derivation of two constants F(0) and f(k) may be difficult.

A method of getting approximate solution to (1.52) is to treat the integral equation as an approximate simultaneous equation for discrete valued y and solve it (see Moran (1954), (1955), (1958)). The solution for (1.52) has been derived for a few different forms of input and output: These are given in the following cases. The solutions given below for continuous Z_t apply to waiting time, storage level either in a dam or an inventory system.

(i) M/G/ and GI/M/1 Systems

In principle, for finite Z_t $(0 \leq Z_t \leq k)$ is we can derive a solution for an M/G/1 system, we derive a solution automatically for a GI/M/1 system by drawing on the properties of the duality principle (which is described in detail in Chapter 2). A dual system is one in which the input and the output process of the primal system is reversed; thus M/G/1 and GI/M/1 systems are dual to each other. If $F_1(y)$ be the d.f. of Z^* in the dual system $\langle Z^*, -u, k\rangle$ and $F(y)$ the d.f. of Z in the primal system, then (see Finch (1960), Ghosal (1968))

$$F_1(y) = 1 - F(k - y) \qquad (0 \leq y \leq k). \tag{1.54}$$

It will be seen in Chapter 2 that the relationship (1.54) can be expressed in terms of the first passage probability V(y), the probability that given the level is y, the system goes empty before filling completely. (See equations (2.4) and (2.14) in Chapter 2). The usefulness of the duality results in the context of M/G/1 and GI/M/1 systems is that if the result of one system is known that for the other system can be immediately derived from (1.54).

Finch (1960) derived the limiting d.f. of Z_t in an M/G/1 system in which Y_t (inter-arrival time/release) follows an exponential d.f. $\lambda \exp(-\lambda x)dx$ $(0 \leq x < \infty, \ \mu > 0)$ and X_t (service time/input) follows a general d.f. $G_1(x)$. He showed that if F(y) (more appropriately F(y,k)) is the solution to (1.52) for such a system, then it is given by

$$F(y) = [\lambda M(y) - M'(y)][\lambda M(k)]^{-1} \qquad (0 \leq y < k) \tag{1.55}$$

where M(x) is a d.f. whose L.S. transform is given by

$$m(\theta) = \int_0^\infty e^{-\theta x} M(x)dx,$$

$$g_1(\theta) = \int_0^\infty e^{-\theta x} dG_1(x)$$

and

$$m(\theta) = \frac{\theta-\lambda + \mu\{g_1(\theta) - g_1(\lambda)\}}{(\theta-\lambda)\ \{\mu g_1(\theta) + \theta - \lambda\}} \tag{1.56}$$

Proof of (1.55). The solution F(y) in (1.55) is a function of two variables k and y. If F(y), or more clearly F(y,k), is of the form $F(y,k) = F_1(k)F_2(y)$, and we try a solution of this form in (1.52) with the specified inter-arrival distribution we can determine $F_1(k)$ and $F_2(y)$. Since the solution F(y,k) is unique, we substittue (1.55) in (1.52) and determine the unknown function M(y).

Let $N(y) = \lambda M(y) - M'(y)$, then after performing the substitution we get

$$N(y) - \int_0^y N(y-z)dG(z) + \lambda g_1(\lambda) \int_0^y N(y-z)e^{\lambda z}dz$$

$$= \lambda g_1(\lambda)e^{\lambda y}\ M(k)\exp(-\lambda k) + \int_0^k M(z)\exp(-\lambda z)dz$$

$$= \lambda g_1(\lambda)M(0)\exp(\lambda y) \tag{1.57}$$

We introduce the L.S. transforms

$$n(\theta) = \int_0^\infty \exp(-\theta x)N(x)dx,\quad G^+(\theta) = \int_0^\infty \exp(-\theta x)dG(x),$$

then from (1.57)

$$n(\theta)[1 - G^+(\theta) + \lambda g_1(\theta)/(\theta-\lambda)] = \{\lambda g_1(\lambda)/(\theta-\lambda)\}M(0);$$

since we have $G^+(\theta) = \lambda\{g_1(\lambda) - g_1(\theta)\}M(0)/\{\theta - \lambda + \lambda g_1(\theta)\}$, we get

$$m(\theta) = \{M(0) - n(\theta)\}/(\theta-\lambda); \tag{1.58}$$

we can take $M(0) = 1$. The equation (1.56) follows from (1.58). This treatment is due to Finch (1960,pp46-47)].

Once we determine the d.f. of Z_t for an M/G/1 system, we can obtain the d.f. of Z_t in a GI/M/1 system with X_t following the exponetial d.f. $\lambda\exp(-\lambda x)dx(\lambda > 0; \quad 0 \leqslant x < \infty)$ from the duality relation (1.54).

(ii) <u>$D/E_p/1$ System</u>

A constant (deterministic) value of Y_t (release) is come across in the model of a dam which is subject to a random input (X_t an r.v). In the context of a queue, this implies that the inter-arrival time is constant by a control system whereas the service time is a random variable: this situation is come across in many production processes (e.g. in a steel plant when ingots are moved to blooming mill, etc). Prabhu (1958a) gave an elegant method of determining the p.d.f. of Z_t when the capacity of the storage system is finite, and the input X_t follows a gamma distribution

$$dG(x) = \frac{\lambda^p}{(p-1)!} e^{-\lambda p} x^{p-1} dx \quad (0 \leqslant x < \infty; \ p \geqslant 1) \qquad (1.59)$$

(in the usual definition of Erlang's d.f. E_p we write λp for λ in (1.59) and $Y_t = m$ (constant). He derived the d.f. of the storage content $Z_t + X_t$ (just after input in a discrete time process), defining

$$H(y) = Pr\{Z_t + X_t \leqslant y\} \qquad (1.60)$$

The argument in favour of deriving the d.f. H(y) as against the d.f. F(y) of Z_t lies in the fact that whatever be the limitation of $Z_t (0 \leqslant Z_t \leqslant k)$, $Z_t + X_t$ lies within the range $(0,\infty)$ (see also Moran (1954 and 1958)). Thus the point probabilities of Z_t at 0 and k are given by

$$F(0) = H(m), \ F(k-m) = 1 - H(k) \qquad (1.61)$$

Our basic model (1.2) for such a system is $Z_{t+1} = \max[Z_t + X_t - m,\ k - m]^+$, $0 \leqslant Z_t \leqslant k$. We get the integral equation for H(y) in the following form:

$$H(y) = \begin{cases} \displaystyle\int_m^{m+y} H(x)\,dG(m+y-x), & y < k - m, \\ \displaystyle G(y-k+m) - \int_m^{k} H(x)\,dG(m+y-x), & y \geqslant k - m. \end{cases} \qquad (1.62)$$

We can write the integral equation for H(y) for $k = \infty$ from (1.58). To solve for H(y), when G(x) is of gamma type (1.58), Prabhu (1958a) applied the transformation

$$\phi(y) = e^{-\lambda y}\{1 - H(k-y)\}; \qquad (1.63)$$

it may be observed that the transformation (1.63) is not a general one like L.S. transform: it is valid only for the input d.f. belonging to gamma family, because the parameter λ comes up in the transformation. Applying (1.63) in (1.62), we obtain after simplification:

$$\phi(y) = e^{-\lambda k}\sum_{r=0}^{p-1} \frac{\lambda^r (k-y)^r}{r!} + \begin{cases} \displaystyle\lambda^p e^{-\lambda m}\int_0^{k-m} \phi(x)(x-y+m)^{p-1}/(p-1)!\,dx & (y \leqslant m), \\ \displaystyle\lambda^p e^{-\lambda m}\int_{y-m}^{k-m} \phi(x)(x-y+m)^{p-1}/(p-1)!\,dx & m < y \leqslant k. \end{cases} \qquad (1.64)$$

If in (1.64) we put

$$\mu = (-1)^{p-1}\lambda^p e^{-\lambda m}, \qquad (1.65)$$

$$\alpha_r = e^{-\lambda k}(-\lambda)^r \sum_{x=0}^{p-r-1} \frac{(\lambda k)^s}{s!} + \lambda^p e^{-\lambda m}(-1)^r \int_0^{k-m} \phi(x)\,\frac{(x+m)^{p-r-1}}{(p-r-1)!}\,dx, \quad r = 0,\dots,p-1. \qquad (1.66)$$

then we get

$$\phi(y) = \begin{cases} \sum_{r=1}^{p-1} \alpha_r y^r/r\,! & (y \leq m), \\ \sum_{r=o}^{p-1} \alpha_r y^r/r! - \int_o^{y-m} \phi(x)(y-m-x)^{p-1}/(p-1)!dx & (m<y\leq k). \end{cases} \tag{1.67}$$

We consider $\phi(y)$ to be analytic within strips of width m: with the help of (1.67) we solve for $\phi(y)$ in the interval $(nm,(n+1)m]$, $n = 1,\ldots,N$, where $k = (N+1)m + u$, $0 < u < m$, N an integer ≥ 1. The solution for (1.63) comes out as

$$\phi(y) = \begin{cases} \sum_{r=o}^{p-1} \alpha_r y^r/r! & (y \leq m), \\ \sum_{r=o}^{p-1} \alpha_r \sum_{q=o}^{n} (-\mu)^q (y-qm)^{qp+r}/(qp+r)! & (nm<y\leq(n+1)m) \end{cases} \tag{1.68}$$

The equations(1.68) and (1.67) give p equations for p unknown quantities α_r, $r = 0,1,\ldots,p-1$. The method of solution is iterative: we get the solution for $\phi(y)$ in the range $(0,m]$, then we get the solution for $\phi(\cdot)$ in the range $(m,2m]$, etc. Hence applying the inverse transformation (1.63) we determine H(y) uniquely. A limitation of this method is that we get solution for H(y) in strips of length m; thus if $m = 1$, we get a lattice distribution for H(y) at points $y = 1,2,\ldots$ etc. With the help of a computer it is not difficult to get the actual numeric solution for H(y). Finally we get the solution for F(y) from (1.63).

<u>Example</u> When $p = 1$,

$$dG(x) = \lambda \exp(-\lambda x)dx \qquad (0 \leq x < \infty); \tag{1.69}$$

from (1.65) and (1.66) we get

$\mu = \lambda \exp(-\lambda m)$,

$$\alpha = \exp(-\lambda k)\sum_{s=o}^{p-1}(\lambda k)^s/s! + \mu\int_o^{k-m}\phi(x)\ (x+m)^{p-1}/p-1)!dx$$

(there is only one α_r, because $r = 1$ as $p = 1$),

and (1.68) leads to

$$\phi(y) = \begin{cases} \alpha & (y \leq m), \\ \alpha \sum_{q=o}^{n}(-\mu)^q(y-qm)^{q+1}/(q+1)! & (nm<y\leq(n+1)m), \end{cases} \qquad (1.70)$$

and hence determineH(y), F(y).

[Remarks: Though we determine H(y), the d.f. of $Z_t + X_t$ by Prabhu's method above, there is no reason why we cannot determine F(y) directly, because the integral equation for F(y) is similar to that H(y) given in (1.62). The reader can also verify that we can also solve (1.62) if we apply the tranformation $\phi(y) = e^{-\lambda y}H(y)$ instead of (1.63)].

(iii) E_p/D/1 System

This system implies that the service time (input) is constant, say m, and the interarrival time (or the release) follows a gamma d.f. (1.59). It may be remarked here that when p tends to infinity E_p tends to D; in other words the release (in this case) becomes constant (= E(Y) = λ, if we write λp for λ in (1.59)); this is the reason why Erlang used λp for λ in (1.59). Thus in model (1.2) we get $Z_{t+1} = \min[Z_t + m - Y_t, k]^+$. The limiting d.f. F(y) of Z_t was obtained by Ghosal (1963a) by applynng Prabhu's method. The integral equation for F(y) is given by (where $G(x) = pr(Y_t \leq x)$):

$$F(y) = 1 - G(m+k-y) + \begin{cases} \int_{y-m}^{k} F(t)dG(m+t-y) & (m<y<k), \\ \int_{o}^{k} F(t)dG(m+t-y) & (y\leq m) \end{cases} \qquad (1.71)$$

when G(x) is of the form (1.59) we get

$$F(y) = e^{-\lambda(m+k-y)}\sum_{r=o}^{p-1} \frac{\lambda(m+k-y)^r}{r!}$$

$$+ \begin{cases} \frac{\lambda^p}{(p-1)!}\int_{y-m}^{k} F(t)e^{-\lambda(m+t-y)}(m+t-y)^{p-1}dt \\ \qquad (m<y<k), \\ \frac{\lambda^p}{(p-1)!}\int_{o}^{k} (\quad " \qquad " \quad)\ dt \\ \qquad (y\leq m) \end{cases} \qquad (1.72)$$

Applying the transformation

$$\phi(y) = F(y)e^{-\lambda y} \qquad (1.73)$$

we get after simplification

$$\phi(y) = \begin{cases} \sum_{r=o}^{p-1} \alpha_r(-\lambda y)^r/r! & (y\leq m) \\ \sum_{r=o}^{p-1} {}_r(-\lambda y)^r/r! - (\lambda^p e^{-\lambda m}/(p-1) \int_{o}^{y-m} \phi(t) \\ \qquad \times (m+t-y)^{p-1}dt & (y>m) \end{cases} \qquad (1.74)$$

where

$$\alpha_r = e^{-\lambda(m+k)}(-\lambda)^r \sum_{s=o}^{p-r-1} \frac{\{\lambda(m+k)\}^s}{s!}$$

$$+ \{\lambda^p e^{-\lambda m}/(p-1)!\}(-1)^r\int_{o}^{k} \phi(t)(m+t)^{p-r-1}dt \ (r = 0,1,\ldots,p-1); \qquad (1.75)$$

finally we get the solution for y in the range [nm,(n-1)m)

$$\phi(y) = \sum_{r=o}^{p-1} {}_{r} \sum_{s=o}^{n} e^{-\lambda sm}\{-\lambda(y-sm)\}^{r+sp}/(r+sp)!; \qquad (1.76)$$

then with the help of (1.73) get the solution to F(y).

In the case of finite k, however, with the help of duality relationship (1.54) [for further remarks see Chapter 2], we can derive F(y) for a system E_p/D/1 and vice versa. Prabhu (1962) solved for F(y) for $k = \infty$ in (1.72) and proved that α_r's could be uniquely determined for such a process.

The merit of Prabhu's method of solution is that it gives an exact solution for F(y) instead of giving an L.S. transform for F(y); the drawback of this method is that computationally it is cumbersome - in actual calculation (through computer) we get awkward solution if $E(u_t) = E(X_t - Y_t)$ is large (+) or too small (-), even for finite k. For continuous Z_t, however, the advantage of this method is that it gives a numerical solution; for discrete Z_t (which is treated subsequently) there is no bigger advantage in computing the solution over the actual solution of the simultaneous equation (analogous to the integral equation (1.62) or (1.71)).

(iv) Given D.F. of u_t

If we do not know the individual d.f.'s of X_t and Y_t, but we know that u_t follows a gamma distribution, then we can solve for F(y) directly from (1.53). As a simple case suppose u_t follows expoenntial distribution:

$$G(x) = pr(u_t \leqslant x) = 1 - e^{-\lambda x} \qquad (0 \leqslant x < \infty; \quad \lambda > 0), \qquad (1.77)$$

then (1.53) reduces to

$$F(y) = 1 - e^{-\lambda(y-k)} - \lambda\int_0^k F(x)e^{-\lambda(y-x)}dx \quad (0 < y < k) \tag{1.78}$$

with the boundary conditions $F(0) = 0$, $F(k+0) = 1$.

Applying the transformation

$$\phi(y) = F(y)e^{\lambda y} \tag{1.79}$$

we get

$$\phi(y) = e^{\lambda y} - e^{\lambda k} - \lambda\int_0^k \phi(x)dx$$

or

$$\phi(y) = e^{\lambda y} - \alpha \tag{1.80}$$

where $\alpha = \exp(\lambda k) = \lambda\int_0^k (x)dx.$ From (1.80) we get

$$F(y) = 1 - \alpha e^{-\lambda y}; \qquad (0 < y < k) \tag{1.81}$$

in other words Z_t follows a modified exponential distribution, in which α can be uniquely determined.

When u_t follows a gamma distribution with parameter p:

$$dG(x) = \frac{\lambda^p}{(p-1)!} e^{\lambda x} x^{p-1} dx, \quad 0 \leq x < \infty, \quad \lambda > 1 \tag{1.82}$$

then (1.53) reduces to

$$F(y) = 1 - e^{-\lambda(y-k)}\sum_{r=0}^{p-1} \frac{\{\lambda(y-k)\}^r}{r!} - \frac{\lambda^p}{(p-1)!}\int_0^k e^{-\lambda(y-x)}(y-x)^{p-1}F(x)dx;$$

so that applying the transformation (1.79) we get

$$\phi(y) = e^{y} - e^{\lambda k}\sum_{r=0}^{p-1} \{\lambda(y-k)\}^r/r!$$

$$- (\lambda^p/(p-1)!)\int_0^k (y-x)^{p-1}\phi(x)dx,$$

which we can express as

$$\phi(y) = \exp(\lambda y) - \sum_{r=o}^{p-1} \alpha_r y^r/r! \qquad (1.83)$$

by suitably defining constants α_i $(i = 0,\ldots,p-1)$ in the way it was done in (1.66). The problem of determining α_i's for $p > 1$ may be analytically difficulty; for the case $p = 1$ however we determine by assuming $F(y)$ to be a distribution function. The approach given here is practically the same as Prabhu's method in a situation in which $m = 0$.

From practical standpoint, however, it would be useful to give a general solution for the integral equation (1.52) where u_t follows roughly a symmetrical general distribution. A gamma distribution for u_t for large value of p in (1.82) or a roughly normal distribution may be more likely. There is scope for doing further work in this topic. In the practical experiences of the author, it may not be very much out of way to make an assumption that u_t follows approximately a normal distribution with a mean nearly zero - a negative $E(u_t)$ would suggest that even when k tends to infinity it may be possible to find a solution. In practical Operational Research (O.R.) work, we resort to simulation when we cannot solve for $F(y)$ analytically, (see Ghosal (1969)); the results obtained therefore enable us to solve numerically various decision-making problems partaining to the system.

1.11 Discrete Models (Z_t following a discrete d.f.)

In Sections 1.8, 1.9 and 1.10 we have principally considered situations in which Z_t (the stock-level, dam level, waiting times, etc) follows a continuous d.f.. In this section we shall consider the situation in which Z_t (stock-level, queue size, etc) follows a discrete (or more generally a lattice) distribution; this arises due to the fact that u_t is discrete ($u_t = \ldots -1, 0,1,2,\ldots$). When we consider the d.f. of a queue size, it is discrete; similarly when we consider the stock-level of cars in a shop, equipments in a factory, etc. we consider the discrete d.f. . The basic approach in determining $F_i(= pr(Z_t \leqslant i))$ as given by (1.7) is basically the same as the one followed in the continuous case: for example, a set of linear equations take the place of integral equations given in Sections 1.8, 1.9 and 1.10. For example the discrete analogue for Lindley's model (for $k = \infty$) is given by (in the limiting case)

$$F_i = \sum_{j \leqslant i} F_j g_{i-j} \qquad (i = 0,1,...) \tag{1.84}$$

where $g_j = pr(u_t = j)$. Even if $0 \leqslant i < \infty$, we can theoretically solve the set of simultaneous equations given by (1.84), if they are not homogeneous; in other words if for any ith equation we get the relationship in the form

$$F_i = c + \sum_{j \leqslant i} F_j g_{i-j} \tag{1.85}$$

then we get a unique solution to the set $\{F_i\}$. For finite k, it is easier to visualize that (1.85) leads to a unique solution.

When the model is of the form (1.1) or (1.2) depending on whether k is infinite or finite, it can be easily realized (see Moran (1958), Prabhu (1965)) that when X_t or Y_t or both are r.v.'s, $\{Z_t\}$ is a Markov chain in most of the situations come across in problems of dam or inventory management; besides $u_t(= X_t - Y_t)$ are mutually independent so that we can write the d.f. F_i of Z_t in the form of a convolution given in (1.84) or (1.85). A bulk service queueing problem (see Bailey (1954)) is the exact prototype of the model of a dam, and can also be treated in the same general fashion. However, when we consider Z_t as the queue size (excluding the person being served) on the departure of the t-th customere, we have for a GI/G/1 queue

$$Z_{t+1} = [Z_t + X_t - 1]^+ . \tag{1.86}$$

In (1.86) $u_t = X_t - 1$, and it can be shown that the series $\{u_t\}$ are mutually independent only if X_t, the number of arrivals during the inter-departure interval, follows a Poisson distribution; in other words we can resort to the technique (1.84) only for an M/G/1 - queue (see Kingman (1966), p.314). For the same reason, the difference-differential equation technique for deriving the queue size distribution for M/M/s or $M/E_p/s$ queues (see Prabhu (1965)) is not applicable for GI/G/1 queues. A possible approach is to relate the queue size distribution to the distribution of waiting time, service time and inter-arrival time (see Finch (1959), Ghosal (1968)). We shall deal with this aspect in greater detail subsequently. We may observe, however, that in the case of a queueing problem, the waiting time process $\{w_t\}$ and the queueing process $\{Q_t\}$ (where w_t is the waiting time of the t-th customer and Q_t the queue size left by the t-th departure) should not be considered separately but the joint process $\{w_t, Q_t\}$ gives us a clear picture of the process. Thus if for any system $\langle\cdot\rangle$ the d.f. of w_t exists, the d.f. of Q_t also exists. This aspect has also been considered in greater detail in the discussion on associated random walks (Chapter 2).

(i) A Generating Function Method for a Simple Dam

We can write the model of a dam (1.2) with finite capacity k with release $Y_t = 1$ at the end of every time period as follows:

$$Z_{t+1} = \min[k, Z_t + X_t - 1]^+; \tag{1.87}$$

we may also interpret that we measure Z_t and X_t in units of the amount of release which has been pre-fixed. Though the model (1.87) with Z_t following a discrete d.f. is of academic interest only (because the volume of water is physically not an integer or multiple of an integer), the method of deriving the limiting d.f. of Z_t, given the d.f. of X_t, is useful in a few queueing or inventory models. Let $f_i = pr(Z_t = i)$, $g_i = pr(X_t = i)$, $i = 0,1,\ldots$; then on the assumption that $\{Z_t\}$ is a Markov chain, we get (see Prabhu (1958b, 1965)):

$$f_o = (g_o + g_1)f_o + g_o f_1,$$
$$f_1 = g_2 f_o + g_1 f_1 + g_o f_2$$
$$\vdots \qquad (1.88)$$
$$f_{k-1} = g_k f_o + g_{k-1} f_1 + \ldots + g_o f_k ,$$

$$f_k = (1-q_k)f_o + (1-q_{k-1})f_1 + \ldots + (1-g_o)f_k$$

where $q_i = \sum_{j=o}^{i} g_j$. Let $v_i = f_i/f_o$; we get the unique solution for the generating function (g.f.) $V(z)$ of $\{v\}$ in terms of the g.f. $G(x)$ of $\{g\}$, where

$$V(z) = \sum_{i=o}^{k-1} z^i v_i , \qquad (1.89)$$

$$G(z) = \sum_{i=o}^{\infty} z^i g_i . \qquad (1.90)$$

By multiplying the corresponding equation for f_i in (1.88) by z^i and adding, we can easily prove that

$$V(z) = \frac{g_o(1-z)}{G(z) - z} . \qquad (1.91)$$

If, for example the input $\{X_t\}$ follows a geometric distribution

$$g_i = ab^i \qquad (a = 1 - b,\ i = 0,1,\ldots), \qquad (1.92)$$

then (1.91) reduces to ($\rho = b/a$ is the mean input)

$$V(z) = \frac{a(1-z)}{a(1-bz)^{-1} - z}$$
$$= (1-bz)(1-\rho z)^{-1}$$
$$= (1-bz)\sum_{i=o}^{\infty} \rho^i z^i ,$$

so that

$$v_i = \rho^i - b\rho^{i-1} .$$

Since $\sum_{i=o}^{k-1} v_i = a(1-\rho^{k+1})/(1-\rho)$, we get

$$f_o = (1-\rho)/a\ (1-\rho^{k+1}),\ f_i = \rho^{i+1}(1-\rho)/1 - \rho^{k+1} \qquad (1.93)$$

Thus {f} is also a geometric distribution ; for $k = \infty$, we get the result easily.

The equation (1.91) gives a method for deriving the g.f. of {f} from the knowledge of {g} ; sometimes difficulty arises in getting the inverse of V(z). This is however a main obstacle in the theory of queues and storage systems, most of the results in which have been derived in terms of Laplace transforms. Attempts are however being made to raise the "Laplace Curtain". Thus even for simple d.f.'s of X_t like Negative Binomial or Poisson, the solution through (1.91)becomes fairly involved (Prabhu (1965), pp. 197-200). A direct approach, applicable only to the negative binomial family of input distributions, is given in the following.

(ii) Negative Binomial Input

A method of transformation was developed by Ghosal (1960, 1962b) deriving the d.f. of Z_t in a finite system with any integral m, so that (1.2) gives

$$Z_{t+1} = \min[k-m, Z_t + X_t - m]^+ ,$$

for X_t following a negative binomial distribution

$$g_j = \binom{p+j-1}{j} a^p b^j \ (a=1-b,\ a > 0,\ j = 0,1,\ldots,p > 1) \qquad (1.94)$$

When p = 1, (1.94) reduces to geometric distribution (1.92). Let $F_i = \sum_{j=o}^{i} f_j$; we write an equation of F_i in terms of F_j and g_j: the method we use here is a discrete analogue for Prabhu's method for determining the d.f. of Z_t when X_t follows a gamma distribution (Section 1.10 (ii)). We get

$$F_i = \begin{cases} \sum_{j=o}^{i+m} g_{i+m-j} F_j & (i<k-2m), \\ q_{i+2m-k} + \sum_{j=o}^{k-m-1} g_{i+m-j} F_j & (i \geq k-2m), \\ (i = 0,1,\ldots,k-m-1). \end{cases} \tag{1.95}$$

where $q_i = \sum_{j=o}^{i} g_j$, $F_i = 0$ for $i < 0$ and $F_i = 1$ for $i \geq k - m$.

Substituting (1.94) for g_j in (1.95) we get

$$F_i = \begin{cases} a^p \sum_{j=o}^{i+m} \binom{p+i+m-j-1}{p-1} b^{i+m-j} F_j & (i<k-2m), \\ a^p \sum_{j=o}^{i+2m-1} \binom{p+j-1}{p-1} b^j + a^p \sum_{j=o}^{k-m-1} \binom{p+i+m-j-1}{p-1} b^{i+m-j} F_j & (i \geq k-2m). \end{cases}$$

We apply the transformation

$$F_i = 1 - b^{-(k-m-i)} \phi_{k-m-i}, \tag{1.96}$$

so that by applying the roperty

$$a^p \sum_{j=o}^{i-1} \binom{j+p-1}{p-1} b^j = 1 - b^i \sum_{r=o}^{p-1} \binom{i+r-1}{r} a^r, \tag{1.97}$$

we get

$$\phi_i = \begin{cases} b^{k+1} \sum_{r=o}^{p-1} \binom{k-i+r-1}{r} a^r + a^p b^m \sum_{j=1}^{k-m} \binom{j-i+m-p-1}{p-1} \phi_j & (i \leq m), \\ b^{k+1} \sum_{r=o}^{p-1} \binom{k-i+r-1}{r} a^r + a^p b^m \sum_{j=i-m}^{k-m} \binom{j-i+m-p-1}{p-1} \phi_j & (i>m) \end{cases} \tag{1.98}$$

Putting

$$(-1)^r\alpha_r = a^r b^{k+1} \sum_{s=o}^{p-r-1} \binom{k+s-1}{s} a^s + a^p b^m k \sum_{j=1}^{k-m} \binom{j-i+m+p-1}{p-1} \phi_j$$

$$(0 \leq r \leq p-1), \qquad (1.99a)$$

$$\lambda = (-1)^{p-1} a^p b^m, \qquad (1.99b)$$

we get the solution for ϕ_i in the following:

$$\phi_i = \begin{cases} \sum_{r=o}^{p-1} \binom{i+r-1}{r} \alpha_r & (i \leq m), \\ \sum_{r=o}^{p-1} \alpha_r \sum_{q=o}^{n} (-\lambda)^q \binom{i-qm+r-1}{qp+r}, & i \varepsilon (nm,(n+1)m], \end{cases} \qquad (1.100)$$

where $k = (N+1)m + u$, N is an integer > 1, $u \varepsilon [0,1]$.

Substituting values of ϕ_i obtained in (1.100) in (1.99a), we solve for α_i's ($i = 0,\ldots,p-1$) uniquely. (see Ghosal 1962b). Again resubstituting ϕ_i, so solved in 1.100, in (1.96) we get F_i ($i = 0,\ldots,k-m-1$); from the knowledge that $F_{k-m} = 1$, we solve for the density function at point k - m as follows: $f_{k-m} = 1 - F_{k-m-1}$.

In Ghosal (1962b), the solution was given for the d.f. of $\{Z_t + X_t\}$, viz. $H_i = pr(Z_t + X_t \leq i)$. In some cases it is advisable to work with H_i rather than with F_i, because $\{Z_t + X_t\}$ lies within the trange $[0,\infty)$ whereas Z_t lies in the range $[0,k-m]$. The method described above can be regarded as the discrete analogue of Prabhu's method for solving F(y), the d.f. of Z_t when Z_t is a continuous function (see Prabhu (1958a)).

It is easy to see that the transformation approach (1.96) in applied in solving F_i for an inventory model

$$Z_{t+1} = [Z_t + m - Y_t, k]^+ \qquad (1.101)$$

in which the supply at the end of each time period is deterministic (m), while Y_t (the demand) follows a negative binomial distribution. This model is of practical use in dealing with inventory problems with aircraft parts for which the demand follows the negative binomial distribution closely (see Brown (1965)).

A remark about this method is that the equation (1.95) for f_i can be solved directly with the help of an electronic computer; the method of solving ϕ_i (1.100) also involves a good amount of computation. For $p = 1$, $m = 1$, the solution through (1.100) should agree with the solution (1.93).

(iii) Queueing with bulk service

A queueing problem with service being rendered to a bulk of m customers at a time can be represented by the model (1.2)

$$Z_{t+1} = \min[k - m, Z_t + X_t - m]^+$$

where k is the size of the waiting room, Z_t is the number waiting in the queue at time τ_t just after the completion of service of a batch of m customers, X_t the arrivals during the interval between the calling of a batch for service. The interval T_t $(= \tau_{t+1} - \tau_t)$ is usually an r.v. following some d.f. Bailey (1954) considered the problem of determinign the d.f. of $\{Z_t\}$ (in the limiting case). when $k = \infty$; he derived solution for the situation in which X_t follows a Poisson distribution. It may be remarked here that the process $\{Z_t\}$ as defined above is a Markov process imbedded in the process Z(t), the queue size at any time t (see Prabhu (1965)). Let number of arrivals (X_t) during the interval T_t follow a Poisson distribution (dropping suffix t)

$$\Pr(X_t = r) = e^{-\lambda T}(\lambda T)^r/r! \quad (r = 0,1,\ldots); \tag{1.102}$$

if T follows a gamma d.f.

$$f(T)dt = \alpha^p e^{-dT} T^{p-1}/(p-1)!\ dT \quad (\alpha>0,\ p>1,\ 0\leq T<\infty),$$

then

$$\Pr(X_t = r) = \int_0^T e^{-\lambda T}(\lambda T)^r/r!\ f(T)dT$$

$$= \{(r+p-1)!/r!(p-1)!\}\lambda^r \alpha^p/(\lambda+\infty)^{r+p}, \tag{1.103}$$

which is a negative binomial distribution (1.94) with $a = \alpha/(\lambda+\infty)$ and $b = \lambda/(\lambda+\infty)$. Hence the method of solution is identical with the one developed in (ii). Bailey (1954) derived the solution for this model, when $k = \infty$, by the method of generating function.

(iv) The (s,S) Policy

In an (s,S) inventory policy, orders are placed as and when the stock in store, plus any quantity already on order, falls to some fixed level s, so that when such a shortfall occurs the stock is replenished to its capacity S(>s). When Z_t and Y_t (the demand over the t-th time period) are discrete, our model is (see Ghosal (1963a)):

$$Z_{t+1} = \begin{cases} Z_t - Y_t & (s < Z_t < S\ ), \\ S - Y_t & (Z_t \leq s). \end{cases} \tag{1.104}$$

If $f_i = pr(Z_t = i)$, $g_i = pr(Y_t = i)$, then in the limiting case we get

$$f_i = \begin{cases} (\sum_{j=o}^{s} f_j)g_{S-i} + \sum_{j=i}^{S} f_j g_{j-i} & (i > s), \\ (\sum_{j=o}^{s} f_j)g_{S-i} + \sum_{j=o}^{S} f_j g_{j-i} & (i \geq s) \end{cases} \qquad (1.105)$$

Given any general $\{g_j\}$ we can solve for $\{f_i\}$ from the set of simultaneous equations (1.105); for an exponential or negative binomial distribution

$$g_j = \binom{p+j-1}{p-1} a^p b^j \quad (a = 1 - b > 0; \quad j = 0,1,\ldots),$$

we can solve (1.105) by applying the transformation

$$f_i = b^{-i} \phi_i \qquad (1.106)$$

and proceeding along the lines followed in (ii). It may be interesting to observe that we can solve for the case when the demand is back-logged, in which case we assume Z_t to be an integer-valued variable in the range $(-\infty,S]$. The same applies when we consider the case of continuous Z_t (the inventory of oils, gases, etc.).

(v) Method of Difference-Differential Equations

A study of probability distribution of Z_t will be incomplete without considering the method of difference-differential equations, which is perhaps the oldest method discovered and also an extremely powerful technique (see Prabhu (1965) for details).

When we consider an M/M/1 system in which the probability of an arrival during an infinitesimal time interval δt is $\lambda\ \delta t$, and the probability of a customer's completion of service during this interval is $\mu\delta t$, the difference equation for $p_i(t)$, probability that the queue size is i at time t, is given by

$$p_i(t+\delta t) = \lambda\delta t p_{i-1}(t) + \mu\delta t p_{i+1}(t) + (1-\lambda\delta t-\mu\delta t)p_i(t) \qquad (i \geq 1)$$

or with $\delta t \to 0$,

$$dp_i(t)/dt = -(\lambda+\mu)p_i(t) + \lambda p_{i-1}(t) + \mu p_{i+1}(t) \qquad (i \geq 1) \qquad (1.107)$$

and $dp_o/dt = -\lambda p_o(t) + \mu p_1(t)$. (1.108)

The set of difference-differential equations (1.107), (1.108) can be solved by the method of generating functions. When $dp_i(t)/dt = 0$, $p_i(t)$ is the same for all t, and we write Lim $t \to \infty$ $p_i(t) = p_i$, suggesting that a stationary stage has been reached. It is easy to find a solution for $\{p_i\}$, viz.

$$p_i = (1-\rho)\rho^i \qquad (i = 0,1,\ldots) \qquad (1.109)$$

where $\rho = \lambda/\mu < 1$. When the size of the waiting room is limited (k say), a limiting d.f. p_i exists if $\lambda < \infty$ (see Finch 1958), and we get

$$p_i = (1-\rho)\rho^i/(1-\rho^{k+1}), \quad i = 0,1,\ldots k. \qquad (1.110)$$

We can also derive the limiting d.f. p_i for an M/M/s, for which also we get the solution by developing the difference-differential equations of the type (1.107), (1.108) (see Prabhu (1965), Cox and Smith (1961) for further details). Morse (1958) used this method for determining queue size distribution for an $M/E_p/1$ system. When $dp_i(t)dt \neq 0$, we get the solution to the time-dependent probabilities $p_i(t)$.

1.12 Concluding Remarks

In this chapter various simple models in queueing and storage problems have been presented, and the methods of determining the d.f. of Z_t have been discussed. The main aim has all throughout been to give a unified treatment to the problems in queueing and storage problems and approach them from a wider perspective.

CHAPTER 2

FIRST PASSAGE PROBLEMS AND DUALITY RELATIONS

2.1 Random Walk Analogy

The model of a system $\{Z_t, u_t, k\}$ (see Sections 1.1, 1.5 and 1.5), where u_t is a stochastic variable, can be regarded as a random walk of the process $\{Z_t\}$. Suppose a particle traverses a straight line path; at the beginning of each time interval (t, t+1) or (τ_t, τ_{t+1}), where $t = \ldots,-1,0,1,2,\ldots$, the particle is at position Z_t (as measured from the origin 0), and immediately after has a displacement u_t $(-\infty < u_t < \infty)$; then if it does not have any further displacement, the position of the particle at the end of the time period is given by

$$Z_{t+1} = \min\{Z_t + u_t, k\}^+ . \tag{2.1}$$

The above implies that Z_{t+1} cannot be negative, at the same time if $Z_t + u_t$ exceeds k, the particle continues to be at k; thus we have two impenetrable barriers at 0 and k. If $k = \infty$, then we have a walk on a line of infinite length with a single impenetrable barrier at 0. Similarly we can conceive of a walk Z_t which has only one barrier at k but the line runs to infinity in the negative zone; here we get

$$Z_{t+1} = \min\{Z_t + u_t, k\}, \quad -\infty < Z_t \leq k; \tag{2.2}$$

the model corresponds to an inventory system in which backlogging of orders is permitted.

Thus we have a one-dimensional random walk; similarly we can conceive of an n-dimensional walk where $Z_t = (Z_{1t},\ldots,Z_{nt})$, $U_t = (U_{1t},\ldots,u_{nt})$ and $k = k_1,\ldots,k_n)$ are n-vectors: this applies particularly to multi-inventory systems. Here we study the properties for the one-dimensional walks follow in most of the cases. There are however a number of distinct problems for multi-dimensional problems: for example the walk of a particle i may not be independent of the walk of the particle j ($i \neq j$). We may call such a system as one of cybernetic random walks in which the position or displacement of one particle affects the position or displacement of a particle or particles in other dimensions: there is continually feedback effect from one walk to the other. Rises and falls of shares in a share market may be conceived as operations in a complex system subject to cybernetic random walks. A simpler system of cybernetic random walks is generated by the waiting time process (w) and the queue size (Q) together, giving a two-dimensional system of walks $Z(=(w,Q))$.

2.2. Conditions for existence of limiting d.f.'s of Z_t

It is worthwhile reviewing the conditions for existence of the limiting d.f. of Z_t (see Section 1.6). From the work of Lindley (1952) we know that if $k = \infty$, the walk Z_t given by (2.1) has a limiting d.f. $F(y) = \lim t \to \infty \operatorname{pr}(Z_t \leq y)$ if $E(u_t) < 0$; if $E(u_t) \geq 0$, the $\operatorname{pr}(Z_t \leq y)$ diverges. This condition, however, holds true strictly (i) if the walk $\{Z_t\}$ is a one-dimensional walk on a line $[0,\infty)$, and (ii) if the walk $\{Z_t\}$ has no feedback effect from any other walk, (this remark is due to the present author). Again from the work of Finch (1958, 1960) we find that if for the model (2.1) k is finite, the limiting d.f. F(y) exists if $E(|u_t|) < \infty$. Thus we have the following theorem (without proof) for a one-dimensional walk.

Theorem 2.1 (i) If we have a walk Z_t within the range $[0,\infty)$, then Z_t has a limiting d.f. F(y) if $E(u_t) < 0$: this holds true if the walk has no feedback effect from any other random walk.

(ii) If the walk is on a range [0,k] then irrespective of whether the walk has feedback effect from other walks the limiting d.f. f(y) exists if $E(|u_t|) < \infty$.

Let us now consider a system of two associated walks, Z_t being a two-vector $Z_t=(Z_t^{(1)}, Z_t^{(2)})$, and $u_t= (u_t^{(1)}, u_t^{(2)})$. Suppose $Z_t^{(1)}$ be a walk within a finite range [0,k], while $Z_t^{(2)}$ a walk on an infinite range $[0,\infty)$. Let us suppose both $Z_t^{(1)}$ and $Z_t^{(2)}$ have feedback effect on each other. Then, considering the walk $Z_t^{(2)}$ independently of $Z_t^{(1)}$, we find that the limiting d.f. $F^{(2)}(y) = pr\{Z_t^{(2)} \leq y\}$ exists if $E(|u_t^{(2)}|) < 0$, and the limiting d.f. of $Z_t^{(1)}$, considered independently, always exists; but if our control phenomenon is that a walk $Z_t^{(2)}$ exists if a walk $Z_t^{(1)}$ exists, then irrespective of whether $E(|y_t^{(2)}|) < 0$ or ≥ 0, the limiting d.f. of $Z^{(2)}$ always exists. An example is the system $Z_t= (w_t, Q_t)$, where w_t is the waiting time of the t-th customer and Q_t is the queue size left by the t-th customer; if either w_t or Q_t is bounded, no matter what the other is (i.e. either w_t is within a finite range (0,T) or the size of the waiting room is finite ($Q_t \leq N$, N finite) the limiting d.f.'s of both w_t and Q_t exist. Thus, from a heuristic argument we state the following theorem with respect to inter-connected random walks (see Ghosal (1968), p. 167, Theorem 3).

<u>Theorem 2.2.</u> If we have a series of random walks $\{Z_t^{(j)}\}$ (j = 1,...,r), where $Z_t^{(j)}$ is the position of the particle on the j-th walk at time t, $u_t^{(j)}$ the displacement of the particle in the j-th walk at time t (t being discrete, t = ..., -1,0,1,...) and each walk has a feedback effect on the other walks either directly or through another walk, then the system $\{Z_t\}$ (= $\{Z_t^{(1)},\ldots,Z_t^{(r)}\}$) has a limiting d.f. F(y) (= $\{F^{(1)}(y_1),\ldots,F^{(2)}(y_2)\}$) if at least one of the walks is over a finite path.

The above theorem is the basic property of cybernetic walks; there is a great scope of studying these systems more analytically. This topic will not be discussed any further in these lecture notes. In a way of physical explanation of a r-component system Z_t it may be said that when a particular component $Z_t^{(j)}$ reaches the impenetrable barrier $k^{(j)}$ (let us assume so), the inability of this walk to move beyond this barrier exerts a control over an inter-connected walk $Z^{(i)}$ affected by the feedback effect of the j-th walk ($i \neq j$). Thus an artificial barrier is created in the walk $Z^{(i)}$ even though it is on range $[0,\infty)$, so that for all practical purposes it is a walk on a range $[0,k^{(i)}]$, where $k^{(i)}$ is variable, large but finite. The waiting time - queue size two dimensional walk $Z\{= (w,Q)\}$ can also be interpreted this way.

2.3 First Passage Probability

Let us consider the one-dimensional random walk $\{Z_t\}$ over a line over the range $[0,k]$, as defined in Section 2.1. Let us define the following probabilities:

$V(y)$ = probability that, given the initial position of the particle is y $(0 < y < k)$, the walk $\{Z\}$ reaches the barrier 0 before reaching the barrier k;

$W(y)$ = probability that, given the initial position of the particle is y, the process reaches the barrier k before reaching the barrier 0.

The probabilities $V(y)$ and $W(y)$ are in a sense limiting probabilities; they do not take into consideration the time or steps in which the particular barrier is reached - they denote that the particular barrier is reached before reaching the other barrier no matter in whatever time or number of steps. It is easy to see that

$$V(y) = 1 - W(y), \qquad 0 < y < k, \tag{2.3}$$

and $V(0) = 1$, $V(k) = 0$, $W(0) = 0$, $W(k) = 1$. In the contex of a finite dam (or a storage system) $V(y)$ denotes the probability that it empties (given the initial level to by y) before filling completely; in the context of waiting time problems in which customers do not wait beyond a time k, $V(y)$ implies the probability that, given at a time the waiting time of a customer is y, the queue becomes subsequently empty before any customer leaves the queue without service (after waiting for a time k). It is also seen easily that if $k = \infty$, $V(y) = 1$ for all y.

We prove the Theorem 2.3 below giving a relation between the first passage probability $V(y)$ with the limiting d.f. $F(y)$ of Z_t. This relation was first established by the present author in respect of a particular type of input d.f.'s and later generalized as below (see Ghosal 1960, 1964). We assume that Z_t has a continuous d.f. .

Theorem 2.3 The probability function $V(y)$ can be expressed in terms of the d.f. of the position of the particle by the relation

$$V(y) = F(k-y), \qquad 0 \leq y < k. \tag{2.4}$$

Proof. We give a combinatorial proof here: it does not tacitly assume that u_t's are all mutually independent. Since (2.4) is obvious for $y = 0$, we suppose $0 < y < k$ throughout the proof. Let A be the event that the walk starting from y reaches the impenetrable barrier at 0 before reaching k and A_j the event that it does so on the j-th step. Then

$$A = \bigcup_{j=1}^{\infty} A_j.$$

We can express A_j as follows:

$$A_j = \bigcap_{\ell=1}^{j-1} (y + \sum_{r=1}^{\ell} u_r > 0) \bigcap (\sum_{r=1}^{\ell} u_r < k-y) \bigcap (y + \sum_{r=1}^{j} u_r \leq 0)$$

$$= \bigcap_{\ell=1}^{j-1} (y + \sum_{r=1}^{\ell} u_r > 0) \bigcap (\sum_{r=1}^{\ell} u_r < k-y) \bigcap (k + \sum_{r=1}^{j} u_r \leq k - y). \quad (2.5)$$

In (2.5) if we replace u_r everywhere by u_{-r} the probabilities are unaltered; let A_j^1 be the result of this replacement on A_j. Thus

$$A_j^1 \subset \bigcap_{\ell=1}^{j-1} (\sum_{r=1}^{\ell} u_{-r} \leq k-y) \bigcap (k + \sum_{r=1}^{j} u_{-r} \leq k-y) \quad (2.6)$$

The right-hand side (R.H.S.) of (2.6) is the event that a walk Y_t, which has only one impenetrable barrier at 0, starting from $Y_{-j} = k$ has a position Y_o not greater than $k - y$. Since the walk Y_t is always to the right of the walk Z_t, it follows that $A_j^1 \subset \{Y_o \leq k - y\} \subset \{Z_o \leq k - y\}$, and

$$A^1 = \bigcup_{j=1}^{\infty} A_j^1 \subset \{Z_o \leq k - y\}. \quad (2.7)$$

Again we may write

$$\{Z_o \leq k - y\} = \bigcup_{j=1}^{\infty} B_j,$$

where the event B_j is the event $Z_o \leq k - y$ and that Z_{-j} is the last Z before Z_o to equal k. If we consider the walk Y_t defined above and starting from k at time -j, then the event B_j implies that either $Y_o \leq k - y$ or $Y_o \neq Z_o$. But on B_j we must have $Y_o = Z_o$ because the walk Z_t never reaches the value k for $-j < t \leq 0$, so that the barrier at k has no effect. Hence

$$B_j \subset \{Y_o \leq k - y\}$$

$$\subset \bigcap_{\ell=1}^{j-1} \{\sum_{r=1}^{\ell} u_{-r} \; k - y\} \bigcap \{k + \sum_{r=1}^{j} u_{-r} \leq k - y\}$$

(since u_r has a continuous d.f.)

$$\sim \bigcap_{\ell=1}^{j-1} \{y + \sum_{r=1}^{\ell} u_r < k\} \bigcap \{y + \sum_{r=1}^{j} u_r \leq 0\}$$

(by changing u_r into u_{-r}, "$\sim$" standing for equivalence in probability)

The event on the R.H.S. implies that the walk starting at y does not reach k on the first j-1 steps except possibly if it reaches 0 first, and the second event implies that it reaches 0 only at the j-th step. Hence it implies the event $\bigcap_{\ell=1}^{j-1} \{y + \sum_{r=1}^{j-1} u_r > 0)$; hence we get

$$B_j \subset A_j.$$

From (2.6), we get $A^1 \subset \bigcup_{j=1}^{\infty} B_j$ where $A_j^1 \sim A_j$; and since

$$\bigcup_{j=1}^{\infty} B_j \subset \bigcup_{j=1}^{\infty} A_j. \tag{2.8}$$

$$\bigcup_{j=1}^{\infty} B_j \sim \bigcup_{j=1}^{\infty} A_j = A \tag{2.9}$$

so that $$\mathrm{pr}\{\bigcup_{j=1}^{\infty} B_j\} = \mathrm{pr}\{A\},$$

which proves the Theorem 2.3

The result (2.8) can also be interpreted in the context of practical problems. The event A_j implies, in the context of waiting time problems that in a stable queue of GI/G/1 type if the present customer being served waits for a time y $(0 < y < k)$, then the queue will be empty only after serving another j customers. It is easy to see that if the present queue size is greater than j, $\mathrm{pr}(A_j) = 0$. The interpretation of A_j in the context of dam or inventory theory is self-evident. The event B_j implies that if a storage is full now, then after j time periods the level will be below $k - y$.

2.3.1 Remarks on Method given above

It will be seen that the method adopted in Section 2.3 in proving Theorem 2.3 is valid under following conditions:

(i) The variables Z_t and u_t are continuous;

(ii) The d.f. of u_t does not change during the entire operations, and Z_t attains a limiting d.f.

(iii) If we define the partial sums $\sum_{r=T+1}^{T+j} u_r = U(T,j)$,

then $U(T,j) \sim U(j)$ for any T in probability (for the time being we assume r to be discrete).

(iv) It is not necessary, for (iii) to hold true, that u_r's are mutually independent. For example u_1 and u_2 may be correlated, but if $U(2) \sim u_1 + u_2 \sim u_2 + u_3 \sim \ldots$, the method holds true.

If we assume that u_r's are mutually independent and have the d.f. $G(x) = pr(u_r \leqslant x)$, then we can prove (2.3) by the method of integral equations. The integral equation for F(y) for model (2.1) is given by

$$F(y) = G(y-k) - \int_0^k F(t)dG(y-t), \quad 0 \leqslant y < k. \tag{2.10}$$

The integral equation for V(y) is given by

$$V(y) = G(-y) + \int_0^k V(t)dG(t-y), \quad 0 \leqslant y < k. \tag{2.11}$$

If a unique solution for each y exists both for (2.10) and (2.11), then by substituting the relation (2.4) in (2.10) we get (2.11) (see Ghosal (1960), (1963)).

2.4 Equivalent Events

Let us consider the one-dimensional random walk $\langle Z_t, u_t, k\rangle$, as shown in Fig. 2.1, being the

Fig. 2.1

graphical interpretation of the model (2.1). The event A_j, as given in Section 2.3, implies that the walk, starting from y, moves randomly till it reaches the barrier 0 before reaching the other barrier at k in exactly j steps: thus the net effect after j steps is a displacement just exceeding y towards left. The event B_j implies that the walk, starting from the barrier k, suffers a total displacement of length exceeding y towards left in j steps (it does not reach k at any step n, $0 < n \leqslant j$). If $B = \bigcup_{j=1}^{\infty} B_j$, then since $A = \bigcup_{j=1}^{\infty} A_j$

$$A \sim B \tag{2.12}$$

From the above we find that by applying set-theoretic arguments (as in Sec. 2.3) we can construct a number of such equivalent sets. For example let C be the event that a particle starting from 0 attains a position to the right of y without returning to 0, and C_j the corresponding event completed in j steps, then we find

$$C \sim D \tag{2.13}$$

where D is the event that the particle starting from the position k - y reaches the barrier k on the right without reaching the barrier 0. We easily find that

$$\mathrm{pr}(C) = \mathrm{pr}\{Z \geqslant y\}.$$

Next change the scale of the line in Fig. 2.1. Change the barrier k into barrier 0, and measure from 0 towards left, so that the altered barrier at left is now k.

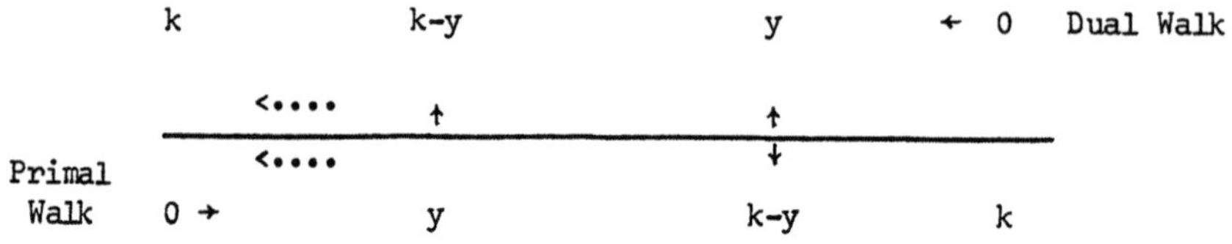

Fig. 2.2

This walk we call the dual walk $\{Z_t^*\}$, while the original walk $\{Z_t\}$ we call the primal walk. The walk $\{Z_t^*\}$ is obtained by changing the sign of the displacement; in other words the displacement u_t^* is related to u_t by the relation $u_t^* = - u_t$. Let us now interpret the events A and B in terms of the dual walk by taking into consideration the barrier and net displacement:

A = event that starting from y on the primal walk the particle reaches 0 before reaching k;

B = event that the particle from 0 on dual walk attains a position on or to the left of y.

Then from (2.12) pr(A) = pr(B); in terms of the dual walk $pr(B) = pr(Z_t^* \geq y) = 1 - F^*(y)$; in terms of the primal walk $pr(A) = V(y)$, hence

$$V(y) = 1 - F^*(y), \quad 0 \leq y \leq k; \tag{2.14}$$

and again

$$F(k-y) = 1 - F^*(y), \quad 0 \leq y \leq k. \tag{2.15}$$

2.5 Dual Queues and Storage Systems

The concept of equivalent events given in Section 2.4 and the results therein lead to all the results in dual queues and storage systems. In literature (Prabhu, 1965) the dual queue is referred to as one obtained by interchanging the inter-arrival time and service time distribution patterns in the primal walk; in storage systems it implies changing input and output d.f.'s. A more ganeral system as given in Section 2.4 is that

$$u_t^* = - u_t$$

which does not necessarily mean that in the models

$\langle Z_t^*, X_t^*, Y_t^*, k\rangle$ and $\langle Z_t, X_t, Y_t, k\rangle$, where $u_t = X_t - Y_t$ and $u_t^* = X_t^* - Y_t^*$,

$$X^* \sim Y_t \qquad \text{and} \qquad Y_t^* \sim X_t,$$

but

$$X^* \sim Y_t + \varepsilon \qquad \text{and} \qquad Y_t^* \sim X_t + \varepsilon,$$

for any ε. The equations (2.14) and (2.15) are the basic duality relations. (For further reference see Ghosal (1968, and an unpublished note)). The result (2.15) was obtained independently by Finch (1960). It may also be verified that the probability of emptiness $V(y)$ in the primal dam equals the probability of overflowing $W^*(k-y)$, given level to be $k-y$, in the dual dam.

2.6 Continuous Time Models

So far we have considered only systems $\langle Z_t, u_t, k\rangle$ in which time t is discrete and Z_t is a continuous variable. It can be shown without difficulty that all the results on equivalent events (Sec. 2.3 and Sec. 2.4) hold true when time t is continuous, (model (1.10)). In section (2.3), we write $A = \bigcup_j A(j)$ where j is continuous on an interval $j \varepsilon (0,\infty)$, and $U(\ell) = \int_0^\ell u(r)dr$.

The coninuous analogue of the set (2.4) is given by

$$A(j) = \bigcap_{\ell=o}^{j-o} (y+U(\ell) > 0) \bigcap_{\ell=o}^{j-o} (U(\ell) < k-y) \bigcap (y+U(j) \leq 0) \qquad (2.16)$$

By following the same method we prove Theorem 2.3 and the results in Section 2.4 when $U(\ell)$ is a separable stochastic process.

It may however be reiterated that continuous time models are relevant only in the case of storage systems, not in case of queues.

2.6 Discrete Time, Discrete Z_t

When Z_t is discrete, let $F_i^{(k)} = pr\{Z_t \leq i\}$, when the capacity of the system is k, $g_i = pr\{u_t = i\}$, i = 0,1,... . Then if u_t's are independent mutually, then we get (Ghosal, 1964):

$$F_i^{(k)} = \sum_{j=-\infty}^{i-k} g_j + \sum_{j=o}^{k-1} F_j^{(k)} g_{i-j} \qquad (0 < i < k); \qquad (2.17)$$

and if $V_i^{(k)}$ be the discrete analogue of the first emptiness probability V(y), we get

$$V_i^{(k)} = \sum_{j=-\infty}^{-i} g_j + \sum_{j=1}^{k-1} V_j^{(k)} g_{j-i} \qquad (0 < i < k). \qquad (2.18)$$

Since both (2.17) and (2.18) have unique solutions, we get the following relationship (Ghosal (1964), eqn. (12)):

$$V_i^{(k+1)} = F_{k-i}^{(k)} \qquad (0 < i < k). \qquad (2.19)$$

Likewise we get the relationship for dual systems also.

2.7 Queues with Finite Waiting Space.

When the size of the waiting space is finite (N), if Z_t is the queue size left by the departing customer in a GI/G/1 queue, X_t the number of arrivals during the service time of the t-th customer, $u_t = X_t - 1$, in the dual queue we get

$Z^*_{t+1} = [\min\{N, Z^*_t + 1\} - X^*_t]^+$ where Z^*_t is the queue size before the t-th arrival epoch, X^*_t the number of departures in the inter-arrival time between t-th and (t+1)th arrival epoch. From (2.19) we get $F^*_i = F_{N-1}$, N being the same for two, and hence (see eqn. (16), p.177 Ghosal (1968)): ($f_i = pr\{Z_t = i\}$, $f^*_i = pr\{Z^*_t = i\}$)

$$f^*_i = f_{N-i} \qquad (0 \leq i \leq N) \tag{2.20}$$

The relation (2.20) was obtained by Keilson (1966) only between transportation variants of systems M/G/1 and GI/M/1 only.

Finch (1959) made an interesting observation: if Z_t be the queue size just after the t-th departure in a GI/G/1 queue ($u_t = X_t - 1$), Z'_t the queue size just before the t-th arrival epoch, then Z_t and Z^{**}_t have the same p.d.f. when $N = \infty$; but this relation holds good for finite N also (see Ghosal (1968)): thus

$$f_i = f^{**}_i \tag{2.21}$$

where $f_i = pr\{Z_t = i\}$, $f^{**}_i = pr\{Z^{**}_t = i\}$.

We also relate the d.f. of the queue size with that of waiting time $\{Z_t\}$ as follows. Consider the model $\langle Z_t, X_t, Y_t \rangle$, where X_t is the service time of the t-th customer, Y_t the inter-arrival time between t-th and (t+1)th customers, and let the size of the waiting room be limited to N (finite or infinite). Let Q_t be the queue size left by the t-th customer, and

$$R_j(t) = pr(Q_t = j);$$

then (Finch (1959), Ghosal (1968)):

$$R_o(t) = pr(Z_t + X_t \leq Y_t), \tag{2.22}$$

$$\sum_{j=i}^{N} R_j(t) = pr(Z_t + X_t > Y_t + Y_{t+1} + \ldots + Y_{t+i-1}). \tag{2.23}$$

If we write $H_i(t) = \sum_{j \geq i} R_j(t)$, $Z_t + X_t = S_t$, $D(t,i) = \sum_{j=o}^{i-1} Y_{t+j}$,

then we write

$$H_i(t) = pr(S_t > D(t,i)), \qquad i \geq 1 \qquad (2.24)$$

The equation (2.24) can also be interpreted in terms of storage theory (Moran 1959) in which S_t is the dam content after the t-th input has been received. When we get a limiting d.f. of S_t, $D(t,i) \sim D(i)$ for any t.

2.8 Busy Period Distribution

If n customers are served in a busy period epoch in a queue, then the length of the busy period (B) is given by

$$B = \sum_{t=1}^{n} X_t \qquad (2.25)$$

where X_t is the service time for the t-th customer. The idle time of the server (I_n) following this busy period sequence is given by (see Chapter 1 - Y_n being the inter-arrival time)

$$I_n = Y_n - X_n \qquad (2.26)$$

Thus in a queueing system we get the busy period sequences $B_1, B_2, \ldots$. Let n_r customers be served during the r-th busy period $B_r (r = 1,2,\ldots)$, then

$$M_r(x) = \sum_{n_r} pr(B_r \leq x | \sum_{j=1}^{n_r-1} u_j > 0, \sum_{j=1}^{n_r} u_j \leq 0)$$

$$= \sum_{n_r} pr(B_r \leq x | U(n_r-1) > 0, U(n_r) \leq 0). \qquad (2.27)$$

If B_r has a limiting d.f. we get $\lim_{r \to \infty} M_r(x) = M(x)$. We can also write (2.27) in the following form

$$M(x) = \sum_{n_r} pr(B_r \leq x | A(n_r)) \tag{2.28}$$

where $A(n_r)$ is the event that a particle, starting from the impenetrable barrier 0 on a random walk (with or without a barrier k on the right) returns to 0 in exactly n_r steps (see Sec. 2.3).

We can also enumerate the busy period $B_r(p)$ exactly after the last customer has left; let his waiting time $W_o = y$; then we define

$$M_r(x,y) = \sum_{n_r} pr(B_r \leq x | A(n_r, y)) \tag{2.29}$$

where $A(n_r, y)$ is the event analogous to event A_j in Sec. 2.3. Prabhu (1965) has obtained the d.f. of $M_r(x,y)$ (he defines it T(y), see p. 73 ibid) for M/G/1 and GI/M/1 systems. There is scope for a general development: Kingman (1968) applied Spitzer's identity to obtain L.T. of M(x,y); it is possible to follow a combinatorial approach also. If we consider the form (2.29) there is scope for a renewal theoretic development.

Prabhu (1965, pp. 145-147) has made use of ladder indices to derive the joint distribution of B_r and n_r in (2.29). Consider the probabilities ($U_r = \sum_{j=1}^{r} u_j$)

$$\begin{aligned} P_n(x) &= pr\{n_r = n,\ B_r \leq x\} \\ &= pr\{U_1 > 0,\ U_2 > 0, \ldots, U_{n-1} > 0,\ U_n \leq 0;\ B_n \leq x\} \\ &= pr\{n \text{ is the first weak ladder index of } (-U_n), \text{ and } B_n \leq x\}; \end{aligned} \tag{2.30}$$

$$\begin{aligned} a_n(x) &= pr(U_n \leq 0;\ B_r \leq x) \\ &= pr\left(\sum_{r=1}^{n} Y_r \geq \sum_{r=1}^{n} X_r;\ \sum_{r=1}^{n} X_r \leq x\right) \\ &= \int_0^x \{1 - A_n(t)\} d\, M_n(t); \end{aligned} \tag{2.31}$$

where $A_n(x) = pr(\sum_{r=1}^{n} Y_r \leqslant x)$, then Prabhu has shown that the joint generating function of (n,x) is given by

$$E(z^n e^{-\theta x}) = 1 - \exp\{- \sum_{n=1}^{\infty} \frac{z^n}{n} \int_o^{\infty} e^{-\theta t}\{1 - A_n(t)\}d\, M_n(t)$$

$$(|z| < 1,\ Re(r) > 0).$$

Hence by letting $z \to 1 - 0$ we get

$$E(e^{-\theta x}) = 1 - \exp\{- \sum_{n-1}^{\infty} \frac{1}{n} \int_o^{\infty} e^{-\theta t}\{1 - A_n)t)\}dM_n(t)\}$$

$$(Re(\theta) > 0) \qquad (2.32)$$

The result (2.32) can be obtained by applying Spitzer's identity also

In practical Operational Research work we generally determine a distribution of n, the number of customers served during a busy period by sampling method; so that if we know the d.f. of X_t (service time), we get the mean and variance of x (the length of busy period) from the following:

$$E(x) = E(\sum_{t=1}^{E(n)} X_t)$$

$$= E(n)E(X_t), \qquad (2.33)$$

$$Var(x) = E(n)Var(X_t), \qquad (2.34)$$

where E(•) stands for "expectation".

CHAPTER 3

CYBERNETIC QUEUEING AND STORAGE SYSTEMS

3.1 A Cybernetic System

Any system comprises a set of elements $A = (a_1,\ldots,a_n)$ which are subject to some specified behaviour pattern; for example, a_i's may be stochastic following any particular d.f. or deterministic or there may be a set of inter-relationships among a_i's etc. All these behaviour patterns can be denoted by the set R-which takes into account any auto-regulatory process, rules, d.f.'s, inter-relations, etc. Thus we denote a system S by $\{A, R\}$ (see Klir and Valach, 1965). From this angle let us review a simple queueing or storage system $\langle Z, X, Y, k\rangle$ or $\langle Z, u, k\rangle$; this system is subject to an input process and subject to some rules gives out an output process - in some cases the input and release processes are fed into the system which, subject to the rules of the system, govern the level of the dam or an inventory system $\{Z\}$. The simple model (1.1) or (1.2) can be regarded as the regulatory process of the system. Thus if Z_t (the level of dam or waiting time) is regarded as an output process, then we can represent the process as follows [$A = (X,Y)$, R = (rules of operation, capacity k, etc.)]:

$$S : \{A, R\} \rightarrow Z. \qquad (3.1)$$

Thus we can have a set of systems $S_1, S_2, \ldots$, such that

$$S_i : \{A_i, R_i\} \rightarrow Z_i \qquad (i = 1,2,\ldots). \qquad (3.2)$$

Again Z_i may be a vector instead of being a single valued variable, $Z_i = (Z_{ii}, Z_{i2},\ldots,Z_{ip})$. Thus in a queueing process we may write $Z = (w,Q)$ where w is the waiting time process and Q is the queue size process. We may be interested in various properties of the output variable, e.g. $F_i(x) = pr(Z_i \leq x)$, $\phi_i(Z_i)$ - some profit or loss function relating to Z_i, etc.

From the definition of the system as in (3.1) we find that we can change a system by changing either A or R or both: e.g.

A unchanged, $R \to R \implies S \to S'$

$A \to A'$, R unchanged $\implies S \to S''$

$A \to A'$, $R \quad R' \implies S \to S^*$

A practical utility of studying these possible structural changes is to ascertain the effect of various changes on Z and any function of Z. For example, one practical problem in queueing and storage systems is to ascertain how much change in the p.d.f. of u_t would not alter the p.d.f. of Z_t. This is somewhat like a sensitivity analysis performed in various Operational Research problems which are primarily concerned with cost or benefit functions of Z.

3.2 Cybernetic outlook of Queueing and Storage Problems

From the above we find that in any system we get a number of elements and also a behaviour pattern R which regulates the system by imposing rules, giving learning mechanism, etc. Starting from the simple queueing and storage model (1.1), we can conceive of all other complex models as cybernetic systems. We offer a few examples of queueing and storage systems that regulate themselves:

(i) A queueing system with a limited waiting room capacity k, in which if a customer on arrival finds the waiting room full, leaves the queue and never returns;

(ii) The storage model analogue of (i) is a system $\langle Z_t, u_t, k \rangle$ with capacity k, so that if the input is high, there is an overflow;

(iii) Dams or inventory systems in series, in which the release from dam 1 is the input to dam 2, the release from dam 2 is the input to dam 3, etc.

(iv) Queueing systems in series in which a customer enters the second queue only after getting his service from the first queue, etc.;

(v) System in which the behaviour of X and Y are not the same throughout, but behaviour of X has feedback effect on the other: in stochastic sense, for example, the d.f. of X_t (service time) may change with the change in the d.f. of Y_t (inter-arrival time), the server becomes more efficient when the arrival rate is high and relaxes when the arrival rate is low, etc. (We can have an analogue for this situation in storage problems);.

(vi) In the context of a dam (or an inventory) the release, though deterministic, changes with the level of the dam;

(vii) There may be a multi-server queue in which the customers always tend to go to the counter where the server is most efficient.

For each system cited above, we can make various studies analytically, viz.

(a) the effect of a change of a rule on the stochastic behaviour of Z_t,

(b) conversely to determine optimum set of rules for running the system economically and at the same time with the least possible risk level for relevant variables in the system. A detailed study of all the models given above and all facets of problems have not been studied in these notes; however a few models have been studied and the attempt has been made to develop an approach to the study of these problems. The method of determining the p.d.f. of Z_t for models (i) and (ii) has been dealt with in Chapter 1.

3.3 Isomorphic Systems

The typical definition of isomorphic systems is as follows (Klir and Valach, 1965, p.110) ($S_i = \{A_i, R_i\}$, $i = 1,2$):

"The systems S_1 and S_2 are isomorphic iff

(i) the elements of the universe A_1 can be mutually uniquely assigned to the elements of A_2 and vice versa;

(ii) the elements of the characteristic R_1 can be mutually uniquely assigned to the elements of the characteristic R_2 in such a manner that to each element of R_1 which may represent an oriented relation between two elements of the universe A_1 we always assign that element of the characteristic R_2 that expresses the same oriented relation between the corresponding pair of elements of the universe A_2".

The isomorphic relation between S_1 and S_2 will be denoted symbolically by

$$(S_1 \text{ i } S_2)$$

The above relation is symmetrical , i.e.

$$(S_1 \text{ i } S_2) \Leftrightarrow (S_2 \text{ i } S_1) \tag{3.3}$$

Isomorphy also possesses the transitive property, i.e.

$$(S_1 \text{ i } S_2) \text{ and } (S_2 \text{ i } S_3) \Rightarrow (S_1 \text{ i } S_3) \tag{3.4}$$

In respect of stochastic systems like those come across in queueing and storage problems we define isomorphism with respect to a specific property, for example, the p.d.f. of Z_t, probability of emptiness, etc. A queueing or a storage system S, $\langle Z, u, k\rangle$, subject to the rule say (1.2), can be represented in the form $\{A,R\}$. The rule given by (1.1) or (1.2) is a characteristic ($\subset R$); so we can define a restricted isomorphism between two systems S_1 and S_2 as follows:

(i) Subject to the same rule of operation (1.1), (1.2) or any other model, two queueing/storage systems S_1 and S_2 are isomorphic iff

$$F_k(x) = F_2(x) \qquad \forall\, x, \tag{3.5}$$

where $F_i(x) = pr(Z_t^{(i)} \leq x)$ for system S_i, $i = 1,2$. Representing S_i by $\langle Z^{(i)}, X^{(i)}, Y^{(i)}; k\rangle$, this implies that given same k and the same rule of operation, even if $X^{(1)}$ and $X^{(2)}$ do not have the same p.d.f. or $Y^{(1)}$ and $Y^{(2)}$ do not have the same p.d.f., the systems S_1 and S_2 are isomorphic iff the relation (3.5) holds true.

(ii) The above definition implies that in the above sense a number of systems $S_i(i = 1,\dots,p)$ are isomorphic to each other iff

$$F_1(x) = F_2(x) = \dots = F_p(x) \qquad \forall\, x \tag{3.6}$$

(iii) We can define a restricted isomorphism between two systems S_1 and S_2 such that, given a constant $c > 0$,

$$F_1(x) = F_2(cx) \qquad \forall\, x, \tag{3.7}$$

The relation implies that if for S_1 the capacity $k_1 = k$, then for S_2 the capacity $k_2 = ck$.

Thus we find that we can define restricted isomorphism with respect to a specific property; in such a case two systems may not be isomorphic with respect to other properties.

Let us consider the simple storage/queueing model (1.1) or (1.2), viz. $Z_{t+1}^{(i)} = [Z_t^{(i)} + u_t^{i}]^+$ or $Z_{t+1}^{(i)} = \min[k, Z_t^{(i)} + u_t^{(i)}]^+$, for each system S_i, $i = 1,2,\ldots$. Let $u_t^{(i)}$ be mutually independent for t's within S_i and let $pr(u_t^{(i)} \leq x) = G_i(x)$ for all t. Then we can prove that the property (3.5), viz. that the limiting probabilities $F_1(x)$ and $F_2(x)$ are the same iff

$$G_1(x) = G_2(x) \qquad \forall\ x\ ; \tag{3.8}$$

similarly the property (3.6) holds good among S_i's ($i = 1,\ldots,p$) iff

$$G_1(x) = \ldots = G_p(x) \qquad \forall\ x. \tag{3.9}$$

Thus we find that isomorphism defined in (3.5) and (3.6) can be expressed in terms of the relation between G_i'$(\cdot)$s. The practical impact of this observation is that $F(\cdot)$ depends entirely on $G(\cdot)$, the p.d.f. of u_t's, not on the p.d.f.'s of X_t's and Y_t's (see Section 1.6 - for models (1.1) and (1.2)).

When u_t's are not mutually independent, the conditions (3.8) may not necessarily imply (3.5); similarly (3.9) may not necessarily imply (3.6).

We can also define restricted isomorphism with respect to time-dependent systems; we consider $F(x,t) = pr\{Z_t \leq x\}$ instead of $F(x)$. Again x may be discrete or continuous. For example, we can define $R_j^{(r)} = pr\{Q_n^{(r)} = j,\ n \to \infty\}$, where $Q_n^{(r)}$ is the queue-size left by the nth customer on departure in the r-th queue system, then we define

$$(Q^{(1)}\ i\ Q^{(2)}) \Longleftrightarrow R_j^{(1)} = R_j^{(2)} \quad \forall\ j. \tag{3.10}$$

The concept of isomorphism can be fruitfully exploited in studying queues/dams in network. As a simple case we can consider such a network as a directed graph = (C,P), where C = $(c_1,\ldots,c_n)$ is the set of nodes C queue points or dam sites) in the graph ζ and P = (p_{ij}), i, j = 1,...,n is the transition matrix such that

$$p_{ij} = 1 \text{ if an oriented line exists from } c_i \text{ to } c_j,$$
$$= 0 \text{ if no such line exists.}$$

The relation $p_{ij} = 1$ can be interpreted in the case of network dams that the release from the i-th dam goes to the j-th dam as input; in the case of queues it implies that on receiving service from the i-th counter a customer goes to the counter j. The above (p_{ij} = 1 or 0) implies a determinate system; when the directed paths are not known, P implies a transition probability matrix in a Markovian system with n nodes.

A simple example of isomorphism in two network of dams (or inventory systems or queues) is given in Figure 3.1

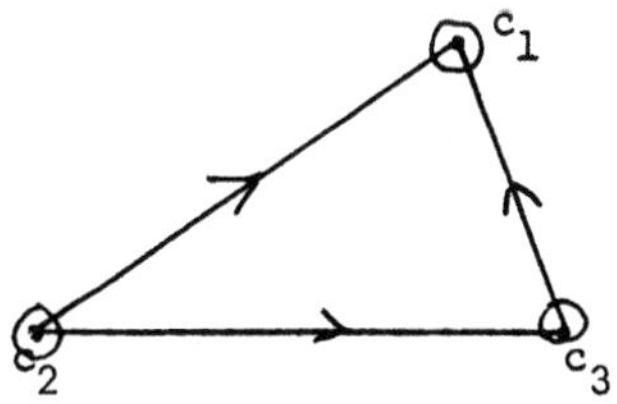

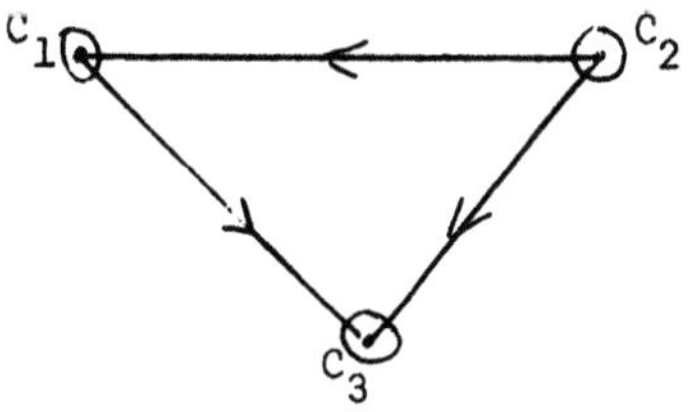

Figure 3.1 Isomorphic Graphs

The graph ζ_1 given by system (i) in Figure 3.1 suggests that there are three dams c_1, c_2 and c_3 in network such that the release from each of dams c_2 and c_3 goes to c_1 while the release from c_2 goes to the dam c_3. The graph ζ_2 given by system (ii) suggests that the dam C_3 gets release from each of dams C_1 and C_2 while the release from the dam C_2 goes to C_1. The two networks ζ_1 and ζ_2 are isomorphic if we assign the elements from sets ζ_1 to ζ_2 as follows:

$$c_1 \longleftrightarrow C_3$$

$$c_2 \longleftrightarrow C_2$$

$$c_3 \longleftrightarrow C_1 \quad .$$

Isomorphism defined above is solely in terms of nodes and their directed graphs. We can elaborate isomorphic behaviour by saying that the releases form X_2 in both have the same p.d.f., the release from c_3 in (i) has the same p.d.f. as C_1 in (ii).

3.4 Queues in Series

An example of a network queueing system is two or more queues in series: in such a system a customer enters queue ζ_1 for service, he goes to the second queue ζ_2 only after getting service in ζ_1, to the third queue ζ_3 only after getting service in ζ_2, etc. The analytical problem is: given the inter-arrival d.f. in ζ_1, and service time d.f.'s in ζ_2, ζ_3,..., how to obtain the (limiting) d.f. of waiting time in various queues. This problem has been attempted by various authors (Jackson (1956), Reich (1957), Ghosal (1962), Saaty (1964), etc.) but the position is that we can obtain the d.f. of waiting times and in different queues only in very simple cases, say (a) when all the queues are of M/M/1 type - which we write M/M/1 → ./M/1 →./M/1 etc., (b) when two queues are of types M/M/ → ./E_2/1 or M/E_2/1 → ./M/1, etc. In most of the practical problems we resort to simulation metho s (an excellent study has been done by Nelson (1966); some studies in steel plants have been done by Tocher (1963)). We briefly discuss the analytical approach (for methods of simulation we refer to Tocher (1963), Ghosal (1969)). Queueing in series is met within most of the production processes, for example in a steel plant.

The analytical approach in studying a system of two queues in series given below is based on the method of integral equations adopted by Lindley (1952) for studying a single GI/G/1 queue. We exploit the concept of isomorphism as follows:

<u>Theorem 3.1</u> If we have two queues, where the waiting time processes $\{Z_t^{(1)}\}$ and $\{Z_t^{(2)}\}$ are given by $Z_{t+1}^{(1)} = [Z_t^{(1)} + u_t^{(1)}]^+$ and $Z_{t+1}^{(2)} = [Z_t^{(2)} + u_t^{(2)}]^+$, where $u_t^{(\cdot)} = X_t^{(\cdot)} - Y_t^{(\cdot)}$ (see Section 1.2), then if the limiting d.f. $F^{(i)}(x) = \text{pr}\{Z_t^{(i)} \leqslant x\}$, i = 1,2, exists, and $G^{(i)}(x) = \text{pr}\{u_t^{(i)} \leqslant x\}$, i = 1,2,

$$F^{(1)}(x) = F^{(2)}(x) \qquad \forall\ x \quad \text{on } [0,\infty) \tag{3.11}$$

iff

$$G^{(1)}(x) = G^{(2)}(x) \qquad \forall\ x \tag{3.12}$$

<u>Proof.</u> The integral equation of $F^{(i)}(x)$ is given by (when u_t's are mutually independent)

$$F^{(i)}(x) = \int_{t \leqslant x} F^{(i)}(x-t)dG^{(i)}(t)\ , \qquad i = 1,2, \tag{3.13}$$

so that if (3.12) holds true, (3.11) holds true and vice versa.

<u>Corollary 3.1.1</u> Thoerem 3.1 holds true if both $Z_t^{(1)}$ and $Z_t^{(2)}$ are over the same finite range [0,k].

<u>Corollary 3.1.2</u> If the d.f. of $u_t^{(i)}$ (= Service time - Inter-arrival time) in the i-th queue in series is the same as the d.f. of $u_t^{(1)}$ in the first queue, then the waiting time d.f.'s of $Z_t^{(i)}$ (i-th queue) and $Z_t^{(1)}$ (first queue) are same, in other words

$$Z^{(i)} \sim Z. \tag{3.14}$$

The approach, therefore, is to determine the d.f. of $u_t^{(i)}$ in the i-th queue and solve for $F^{(i)}(\cdot)$ by Lindley's method, where applicable.

Let us consider two queues in series $-\langle Z_t^{(1)}, X_t^{(1)}, Y_t^{(1)}\rangle$ being the waiting time system for the first queue and $\langle Z_t^{(2)}, X_t^{(2)}, Y_t^{(2)}\rangle$ for the second queue. The waiting time system of the first queue is subject to the model

$$Z_{t+1}^{(1)} = [Z_t^{(1)} + X_t^{(1)} - Y_t^{(1)}]^+$$

while for the second queue in which a customer arrives immediately after service from the first queue is subject to the model

$$Z_{t+1}^{(2)} = [Z_t^{(2)} + X_t^{(2)} - X_{t+1}^{(1)} + X_t^{(1)} + Z_t^{(1)} - Y_t - Z_{t+1}^{(1)}]^+ \tag{3.15}$$

which reduces to

$$Z_{t+1}^{(2)} = \begin{cases} [Z_t^{(2)} + X_t^{(2)} - X_{t+1}^{(1)}]^+ & \text{if } Z_{t+1}^{(1)} > 0, \\ \\ [Z_t^{(2)} + X_t^{(2)} - X_{t+1}^{(1)} + V_t]^+ & \text{if } Z_{t+1}^{(1)} = 0, \end{cases} \tag{3.16}$$

where $V_t = Z_t^{(1)} + X_t^{(1)} - Y_t^{(1)}$, so that $Z_{t+1}^{(1)} = V_t$ if $V_t > 0$ and $Z_{t+1}^{(1)} = 0$ if $V_t \leqslant 0$. Thus in (3.16) $V_t = -I_t$, where I_t is the idle time of the server after departure of the t-th customer (See Section 1.9).

We can write (3.16) as follows:

$$Z_{t+1}^{(2)} = [Z_t^{(2)} + u_t^{(2)}]^+ \tag{3.17}$$

where $u_t^{(2)} = X_t^{(2)} - X_{t+1}^{(1)}$ if $Z_{t+1}^{(1)} > 0$ and $= X_t^{(2)} + X_{t+1}^{(1)} + V$ if $Z_{t+1}^{(1)} = 0$. We can derive the d.f. $G^{(2)}(x)$ of $u_t^{(2)}$ as a mixture distribution if we assume that asymptotically $V_t(= -I_t)$ is independent of $X_t^{(2)} - X_{t+1}^{(1)}$. Thus if $G_1^{(2)}(x) = pr(X_t^{(2)} - X_{t+1}^{(1)} \leqslant x)$ and $G_2^{(2)}(x) = pr(X_t^{(2)} - X_{t+1}^{(1)} + V_t \leqslant x | V_t \leqslant 0)$, then

$$G^{(2)}(x) = pr(V_t > 0)G_1^{(2)}(x) + pr(V_t \leq 0)G_2^{(2)}(x) \tag{3.18}$$

$$= pr(I_t = 0)G_1^{(2)}(x) + pr(I_t > 0)G_2^{(2)}(x) \tag{3.19}$$

$$(0 < x < \infty).$$

(a) M/M/1 → ./M/1

Let Y_t, the interarrival time in the first queue follow the exponential d.f. $\exp(-\mu x)dx$ ($\mu > 0$, $0 \leq x < \infty$), $X_t^{(1)}$, the service time in the first queue, follows the exponential d.f. $\lambda_1 \exp(-\lambda_1 x)dx$ ($\lambda_1 > 0$, $0 \leq x < \infty$), and $X_t^{(2)}$, the service time in the second queue follows the d.f. $\lambda_2 \exp(-\lambda_2 x)dx$ ($\lambda_2 > 0$, $0 \leq x < \infty$). Then we get $pr(V_t > 0) = pr(I_t = 0) = \mu/\lambda_1$ and $pr(V_t \leq 0) = pr(I_t > 0 = 1 - \mu/\lambda_1$ ($\mu/\lambda_1 < 1$ for stability of the system), and

$$G_1^{(2)}(x) = \begin{cases} \lambda_2/(\lambda_1+\lambda_2)\}\exp(\lambda_1 x) & (x < 0), \\ 1 - \{\lambda_1/\lambda_1 + \lambda_2\}\exp(-\lambda_2 x) & (x \geq 0) \end{cases} \tag{3.20}$$

and

$$G_2^{(2)}(x) = \begin{cases} (\lambda_1/\lambda_1 - \mu)(\lambda_2/\lambda_2 + \mu \exp(\mu x) - \lambda_2\mu/\lambda_1(\lambda_1+\lambda_2)\exp(\lambda_1 x)) & (x < 0) \\ 1 - \mu\lambda_1/(\lambda_1+\lambda_2)(\lambda_2+\mu) \exp(-\lambda_2 x) & (x \geq 0) \end{cases} \tag{3.21}$$

whence

$$G_2(x) = \begin{cases} \lambda_2 | (\lambda_2+\mu)\exp(\mu x) & (x < 0) \\ 1 - \mu/(\lambda_2+\mu)\exp(-\lambda_2 x) & (x \geq 0) \end{cases} \qquad (3.22)$$

Thus from (3.22) $u_t^{(2)} \sim (X_t - Y_t)$ where $X_t \wedge \lambda_2 \exp(-\lambda_2 x)dx$

and $Y_t \wedge \mu\exp(-\mu x)dx$ $(\mu,\lambda_2 > 0,\ 0 \leq x < \infty)$. From Theorem 3.1

therefore $Z_t^{(2)}$ has the same d.f. as in a single M/M/1 where interarrival time is the same as the interarrival time in the first queue and the service time as in thr second queue. The property of M/M/1 → ./M/1 queues is that the nature of d.f. of the service time in the first queue (paramter λ_1) is not reflected in the d.f. of $Z_t^{(2)}$. Thus we prove the following theorem.

Theorem 3.2 If we have r Poisson queues in series M/M/1 → ./M/1 → ...→ ./M/1, where the inter-arrival time in the first queue follows exponential distribution with parameter and the service time in the i-th queue follows exponential distribution with parameter λ_i ((i = 1,...,r), then the waiting time in the ith queue follows the same d.f. (exponential) as in a single queue M/M/1 with parameter μ and λ_i.

(This theorem was proved by a different method by Reich (1957), by the above method by Ghosal (1962a)).

(b) M/M/1 → ./G/1

From Theorem 3.2 we derive the following conclusion: if the waiting time in the second queue $Z_t^{(2)}$ has d.f. $F^{(2)}(x)$, $0 \leqslant x < \infty$, the service time $X_t^{(2)}$ has a general d.f., inter-arrival time in the first queue (Y_t) has an exponential d.f. with parameter μ, then

$$Z_t^{(2)} \sim Z_t$$

where Z_t is the single queue system $\langle Z_t, X_t^{(2)}, Y_t \rangle$.

(c) $M/E_2/1$ → ./M/1

Ghosal (1962c) considered such a queueing system in series in which Y_t (the interarrival time in the first queue) follows exponential distribution with parameter μ; $X_t^{(2)}$, the service time in the second queue, follows an exponential distribution qwith parameter λ_2; and $X_t^{(1)}$, the service time in the first queue, follows the gamma distribution $\lambda_1^2 x \exp(-\lambda_1 x)dx$ $(\lambda_1 \leqslant 0,\ 0 \leqslant x < \infty)$. The important result for idle time (I_t) for any M/G/1 was exploited, viz. that (see Section 19)

$$\text{pr}(I_t \geqslant x) = \text{pr}(V_n \leqslant -x) = (1-\mu/E(X_t^{(1)}))\exp(-\mu x), \qquad (3.23)$$

$$= (1-2\mu/\lambda_1)\exp(-\mu x) \quad \text{in the above first queue.}$$

$$(\mu,\lambda_1 > 0,\ 0 \leqslant x < \infty).$$

We get $G_2(x)$ as follows:

$$G_2(x) = \begin{cases} \lambda_1\lambda_2(\lambda_1-2\mu)/(\lambda_1-\mu)^2(\lambda_2+\mu)\exp(\mu x) + \mu\lambda_2/(\lambda_1+\lambda_2)(\lambda_1-\mu) \\ \qquad \times \left\{ \dfrac{\mu\lambda_2+\lambda_1^2}{(\lambda_1+\lambda_2)(\lambda_1-\mu)} - \lambda_1 x \right\} \exp(\lambda_1 x) \quad (x < 0) \\ 1 - \mu\lambda_1(\lambda_1+2\lambda_2)/(\lambda_1+\lambda_2)^2(\lambda_2+\mu)\exp(-\lambda_2 x) \quad (x \geq 0). \end{cases}$$

The characteristic function $\phi_2(\theta) = E(e^{i\theta Z_t^{(2)}})$ was obtained in the following (by using (3.18)):

$$\phi_2(\theta) = (1 - \mu/\lambda_2) \left\{ 1 + \frac{b_o + b_1 i\theta + b_2(i\theta)^2}{c_o + c_1 i\theta + c_2(i\theta)^2 + c_3(i\theta)^3} \right\} \tag{3.24}$$

where b_i's, c_i's $(i = 0,1,\ldots)$ are constants. By inverting $\phi_2(\theta)$ the d.f. $F_2(x)$ of $Z_t^{(2)}$ comes out to be hyper-exponential

$$F_2(x) = \begin{cases} 1 - \mu/\lambda_2 & (x = 0), \\ 1 - \sum_{j=1}^{3} q_j \exp(-d_j x) & (x > 0), \end{cases} \tag{3.25}$$

where q's and d's are constants and $\sum_{j=1}^{3} q_j = \mu/\lambda_2$. In deriving (3.25) the property that the idle time distribution of the server in the second queue, under some approximation, followed the exponential distribution const. $\exp(-\mu x)dx$ $(\mu > 0,\ 0 < x < \infty)$, was used.

3.5 Dams in Series

Let us consider two infinite dams in series; let the release from the first dam is $\min(m_1, Z_t^{(1)} + X_t^{(1)})$, where $X_t^{(1)}$ is its natural input: let this release go as input into the second dam which also has an input $X_t^{(2)}$ from the natural source and let the release from the second dam be $\min(m_2, Z_t^{(2)} + X_t^{(2)} - Y_t^{(2)})$ where $Y_t^{(2)} = X_t^{(2)} + \min(m_1, Z_t^{(1)} + X_t^{(1)})$. Hence we get $Z_t^{(2)}$, the level of the second dam, as the following model:

$$Z_{t+1}^{(2)} = \begin{cases} [Z_t^{(2)} + X_t^{(2)} + m_1 - m_2]^+ \quad \text{if } Z_{t-1}^{(1)} + X_{t-1}^{(1)} > m_1, \\ \\ [Z_t^{(2)} + X_t^{(2)} + Z_{t-1}^{(1)} + X_{t-1}^{(1)} - m_2]^+ \\ \\ \quad \text{if } Z_{t-1}^{(1)} + X_{t-1}^{(1)} \leq m_1, \end{cases} \qquad (3.26)$$

It can be easily seen that the model of $\{Z_t^{(2)}\}$ is not analogous to the waiting time model (3.16) in the second queue in series. The input into the second dam $(X_t^{(2)} + Y_t^{(2)})$ is a mixture of two components, and the d.f. $F^{(2)}(x)$ for $Z_t^{(2)}$ can be obtained through Lindley's method (though the analytical method for deriving the solution for some d.f.'s of $X_t^{(1)}$, $X_t^{(2)}$ may be cumbersome). Nevertheless, the model (3.26) provides a basis for simulation studies for practical problems.

There is scope for studying more complicated problems of queues and dams in network. At the present state of knowledge we bank primarily on simulation methods for solving decision problems with respect to these.

3.6 Adaptive Release Rules

In this section we deal with a few models for storage systems (dams or inventories) for which the release system is made to change with the level Z_t. The release system is mostly deterministic, and the sole objective is to achieve homeostasis by such release rules. The model has been developed particularly with respect to dams and inventories, but applying analogy developed in Chapter 1 we can easily derive results for cybernetic queueing systems (see Ghosal, unpublished paper "Cybernetic Queueing Systems - I"). There is a great scope for developing systems which have stochastic release but at the same time achieve adaptivity.

(i) <u>Rosenblatt's Grain Storage Model</u>

An example of a simple adaptive model is provided by Rosenblatt's model for a grain storage problem (see Rosenblatt, 1954). It is the same as model (1.1) for $\langle Z_t, X_t, Y_t \rangle$ where the release Y_t is a pre-determined proportion of the total storage $(Z_t + X_t)$ after an input X_t has been received. Let $Y_t = (1-\alpha)(Z_t + X_t)$ where $0 < \alpha < 1$. For such a system

$$Z_{t+1} = \alpha(Z_t + X_t), \qquad 0 < Z_t < \infty, \tag{3.27}$$

implying that the question of emptiness does not arise. If initially $Z_o = 0$, then $Z_1 = \alpha X_o$, $Z_2 = \alpha^2 X_o + \alpha X_1$,

and

$$Z_t = \alpha X_{t-1} + \alpha^2 X_{t-2} + \ldots + \alpha^t X_o \; . \tag{3.28}$$

If X_t are identically distributed with $E(X_t) = m$, $var(X_t) = \sigma^2$, then we get

$$E(Z_t) = m \sum_{r=1}^{t} \alpha^r = m\alpha(1-\alpha^t)/(1-\alpha) \tag{3.29}$$

and

$$Var(Z_t) = \sigma^2 \sum_{r=1}^{t} \alpha^{2r} = \alpha^2\alpha^2(1-\alpha^{2t})/(1-\alpha^2). \tag{3.30}$$

For $t \to \infty$, we get $Z_t \sim Z$ where

$$E(Z) = m(1-\alpha), \; Var(Z) = \alpha^2\alpha^2/(1-\alpha^2). \tag{3.31}$$

Rosenblatt considered the cost function (see also Prabhu (1965), p.180)

$$\phi(Z_t) = cZ_t + k_1 + k_2(Y_t - M)^2 \tag{3.32}$$

where c is cost of storage/unit, k_1 and k_2 two constants, M is the desired level of release. The problem is to determine optimal α such that $\phi(Z)$, given c, k_1, k_2 and M, is minimum. $E(Y_t) = m$ for the long run, so that $M = m$, and $E(Y_t - M) = Var(Y_t)$ and

$$E[\phi(Z)] = cm/(1-\alpha) + 2k_2\sigma^2/(1+\alpha); \qquad (3.33)$$

putting $\partial E[\cdot]/\partial\alpha = 0$, we get optimal $(= \alpha_o)$ as follows:

$$\alpha_o = \begin{cases} \sqrt{2k_2\sigma^2/cm} - 1 \,/\, \sqrt{2k_2\sigma^2/cm} + 1 & \text{if } 2k_2\sigma^2 > cm \\ 0 & \text{if } 2k_2\sigma^2 \leq cm \end{cases} \qquad (3.34)$$

Rosenblatt's model may be further sophisticated by giving some rules: e.g. do not release until $Z_t + X_t \geq A$, in the region $A < Z_t + X_t \leq B$, release $\alpha_1(Z_t + X_t)$, in the region (B,∞) release $\alpha_2(Z_t + X_t)$: given $A, B, \alpha_1, m, \sigma^2$ determine optimal α_2 for region (B,∞).

(ii) A release scheme for a finite storage model

A release scheme has been studies by A.C. Trajstman (unpublished, Monash University) for a finite model, $0 \leq Z_t \leq k$. The scheme is as follows:

$$Y_t = (A_t + X_t)^{1+r}/k^r \qquad (0 < r \leq 1). \qquad (3.35)$$

For simple systems, we may assume $r = 1$, in which case the maximum level of Z tends to be at level k/4. The operating rule is: release proportionately more when the stock is high, less when the stock is low - the basic assumption, however, is that $Z_t + X_t$ seldom exceeds k; if X_t is large there is a spillover and for all practical purposes $Z_t + X_t$ comes back to k. Such an inventory policy may be justified for items in heavy demand. With a cost function $\phi(Z)$ of the type (3.32), the optimization problem may be twofold: (a) to determine optimal k such that

$E[\phi(z)]$ is minimum, (b) to determine optimal r such that $E[\phi(z)]$ is minimum ($0 < r \leq 1$). Problems of determining the d.f. of Z_t for such problems are being studied by Trajstman.

(iii) Holdaway's Release Scheme

This is a scheme in which the release from a dam is entirely deterministic, but instead of being kept at a constant level ($Y_t = m$), it is made to change for different levels of the dam storage (Moran, 1955). With the usual definition of Z_t, X_t, Y_t, k, we get the following:

(a) If $Z_t + X_t \geq k + m$, a quantity m is released through the dam and the quantity $Z_t + X_t - k - m$ is spilled over (contributing to flood), so that $Z_{t+1} = k$.

(b) If $\ell + m \leq Z_t + X_t < k + m$, where $\ell < k$, a quantity m is released ($2m < k$, $m < \ell$) and the final state is: $Z_{t+1} = Z_t + X_t - m$.

(c) If $\ell + d \leq Z_t + X_t < \ell + m$, where $d < m$, a quantity $Z_t + X_t - \ell$ is released so that $Z_{t+1} = \ell$.

(d) If $d \leq Z_t + X_t < \ell + d$, a quantity d is released so that $Z_{t+1} = Z_t + X_t - d$.

(e) If $Z_t + X_t < d$, the entire amount $Z_t + X_t$ is released, and $Z_{t+1} = 0$.

The problem of determining the limiting d.f. of $Z_t + X_t$, viz. $H(y) = Pr(Z_t + X_t \leq y)$ was studied by Ghosal (1959) for an exponential input,

$$G(x) = pr(X_t \leq x) = 1 - \exp(-\mu x),\ \mu > 0,\ 0 \leq x < \infty. \tag{3.36}$$

The integral equation for H(y) in terms of H(•) and G(•) is given below:

$$\left.\begin{aligned}
&\text{(a)}\quad \text{for } y < \ell: \\
&\qquad H(y) = -\int_{d}^{d+y} H(t)\,dG(y+d-t), \\
&\text{(b)}\quad \text{for } \ell \leq y < k: \\
&\qquad H(y) = -\int_{d}^{d+y} H(t)\,dG(y+d-t) - \int_{\ell+m}^{y+m} H(t)\,dG(y+m-t), \\
&\text{(c)}\quad \text{for } y \geq k: \\
&\qquad H(y) = -\int_{d}^{d+y} H(t)\,dG(y+d-t) - \int_{\ell+m}^{k+m} H(t)\,dG(y+m-t).
\end{aligned}\right\} \qquad (3.37)$$

We apply the transformation

$$\phi(y) = \exp(-\mu y) \quad \{1-H(k+m-y)\} \qquad (3.38)$$

after substituting (3.36) in (3.37). We follow the method developed in Section 1.10 (ii): we solve for $\phi(y)$ in the ranges $[0,m]$, $(m,k+m-\ell]$, $(k+m-\ell,\infty)$. The range $(m, k+m-\ell]$ is again divided into strips of width m each $(m,2m]$, $(2m,3m]$,... , while the range $(k+m-\ell,\infty)$ is divided into strips of width d each $(k+m-\ell, k+m-\ell+d]$, $(k+m-\ell+d, k+m-\ell+2d]$,... . We put $k = (N+1)m + U$, $\ell = (T+1)m + V$, $0 < U, V < 1$, N and T are integers.

$$\lambda = \mu\exp(-\mu d), \; \nu = \mu\exp(-\mu m),$$

$$\alpha = \exp[-\mu(k+m)] + \nu \int_{o}^{k-\ell} \phi(t)dt + \lambda\int_{k+m-\ell-d}^{k+m-d} \phi(t)dt,$$

$$B = \exp[-\mu(k+m)] + \lambda\int_{k+m-\ell-d}^{k+m-d} \phi(t)dt; \qquad (3.39)$$

then we get in the range $[0,k+m-\ell]$, for the particular strip $(nm,(n+1)m]$,

$$\phi(y) = \alpha \sum_{q=o}^{n} (-\nu)^q (y-qm)^q/q! \ ; \qquad (3.40)$$

in the range $(k+m-\ell,\infty)$ for the strip $(k+m-\ell+nd,\ k+m-\ell+(n+1)d]$, $n = 0,1,2,\ldots$.

$$\phi(y) = \beta \sum_{r=o}^{n} (-\lambda)^r (y-rd-k-m+\ell)^r/r!$$

$$+\alpha \sum_{q=o}^{N-T} (-\nu)^q \sum_{s=1}^{n+1} (-\lambda)^s \frac{\{k-(q-1)m-\ell\}^{q+s}}{(q+s)!} \sum_{r=o}^{n-s} (-\lambda)^r \frac{\{y-(r+s)d-k-m+1\}^r}{r!}$$

$$-\alpha \sum_{q=o}^{N-T-1} (-\nu)^q \sum_{s=1}^{n+1} (-\lambda)^s \frac{\{k-(q-1)m-\ell-d\}^{q+s}}{(q+s)!}$$

$$X \sum_{r=o}^{n-s-1} (-\lambda)^r \frac{(y-(r+s-1)d-k-m)^r}{r!}$$

$$+ (-\lambda)^{n+1} \sum_{q=o}^{N-T-1} (-\nu)^q \frac{\{y-(n+1)d-qm\}^{q+m+1}}{(q+n+1)!} . \qquad (3.41)$$

Hence, solving for α and β, we derive (y), and hence $H(y)$ by applying the transformation (3.38). Hence the d.f. of Z_t is derived as follows:

$$\begin{aligned} &pr\{Z_{t+1} = k\} = 1 - H(k-m), \\ &pr\{\ell < Z_{t+1} < z\} = H(z+m)-H(\ell+m), \quad \ell < z < k, \\ &pr\{Z_{t+1} = \ell\} = H(\ell+m) - H(\ell+d), \\ &pr\{0 < Z_{t+1} < z\} = H(z+d) - H(d), \quad 0 < z < \ell, \\ &pr\{Z_{t+1} = 0\} = H(d). \end{aligned} \qquad (3.42)$$

The detailed method is given in Ghosal (1959).

The operational problem might be to determine a minimal cost function $\phi(z)$ of the type of (3.32) with respect to decision variables d,m,ℓ and k.

Only a few such adaptive release schemes are given above: the idea of such schemes is to decrease the probability of having a large stock by increasing the release when stock is high. There may however be occasional deviation from this principle particularly in the case of dams. For example, if X_t be the input (rainfall in the catchment area) in the t-th time period, then if the water from the dam is to be used entirely for irrigation purposes, we may find that if rainfall exceeds a level say $X_t \geq \alpha$, the adjoining area under irrigation does not require any water from the dam. Under such a system, we may have a scheme: Y_t (release) = m if $X_t < \alpha$ $(\alpha > m)$ and = $m_1 (< m)$ if $X_t \geq \alpha$. In such a case, however, the possibility of flood at high rainfall years is aggravated; to circumvent it we might release $Y_t = m$ when $Z_t + X_t > A$ (a predetermined level), the excess amount is not distributed in farm lands. (Work on such schemes is being done by C.S. Joy, Graduate student in Civil Engineering Dept., Monash University).

3.7 Queueing Problems

The models discussed in Section 3.6 have all analogues in terms of queueing problems; these can be obtained with reference to Table 1.1 in which Z_t can be made to imply either the queue size or the waiting time. Holdaway's scheme (Section 3.6(iii)) has an analogous bulk queueing problem: if the total number of customers including the ones being served $(Z_t + X_t)$ is less that d, serve all of them togther; if they exceed d but are less than $\ell + d$, serve only d together; if they lie between $\ell + d$ and $\ell + m$ serve only $(Z_t + X_t - \ell)$ together; if $(Z_t + X_t)$ lie between $\ell + m$ and $k + m$ together; if $(Z_t + X_t)$ exceed $k + m$, refuse service to $(Z_t + X_t - \ell - d)$.

The bulk queueing problem discussed by Neuts (1967) can also be treated in a similar fashion: it is a modification of Holdaway's scheme: no one is served till $(Z_t + X_t)$ is allowed to grow to d; if $(Z_t + X_t)$ lies between d and m(>d), serve all customers together; if $(Z_t + X_t)$ exceeds m, the first m entire service together, etc.

In single server GI/G/1 queues we occasionally find that the server adapts himself to cope with the traffic: in other words, the distribution of the service time (X_t) is not the same throughout, but changes with the d.f. of the inter-arrival time (Y_t). Thus we may find:

$$X_t \frown P_1(x) \qquad \text{so long} \qquad Y_t \frown R_1(x)$$

$$X_t \frown P_2(x) \qquad \text{"} \quad \text{"} \qquad Y_t \frown R_2(x), \text{ etc.}$$

or $$P_1(x) \rightarrow P_2(x) \Leftrightarrow R_1(x) \rightarrow R_2(x), \tag{3.43}$$

where $P_1(x) = pr(X_t \leq x)$, $R_1(x) = pr(Y_t \leq x)$, etc. The server relaxes when there are fewer customers on the lines, exerts more when the waiting line is long. Thus we have the sensing device mostly through the server's eyes and adaptive device through the change in P(x). Thus we get a self-organization within any such queueing problem.

There is a host of problems in cybernetic systems referred to here, and a great scope to solve them analytically - and where analytical solution do not exist, through computer simulation.

CHAPTER 4

OPTIMAL CAPACITY OF A STORAGE SYSTEM

4.1 The Optimization Problem

We frequently come across the problem of determining the optimal capacity of a storage system with respect to some criteria. For example in a production system which requires adequate storage (or bunker) capacity any method of decision-making for determining the optimal capacity takes into view two conflicting factors, viz. (i) to satisfy the demand for storage when there is an inflow of materials, (ii) to avoid over-investment in building a too large capactiy which may be partially used for sometime. The factor (ii) stated above is particularly tricky, because what we consider too large a capacity today may be inadequate in the next ten years if there has been a considerable growth in production activity. An example is the baffling problem of creating adequate capacity in most of the international airports of the world both from the point of view of meeting the runway capacity for growing airtraffic and at the same time accommodating growing passenger traffic in the airport building. When most of the airports were designed, it was not visualized that huge jumbo jets carrying 400 passengers would be coming. The problem of determining the optimal reservoir capacity of a dam to be built is also similar: the estimates of future demand for irrigation water and hydroelectricity generation may be offset by the development of the area after a few years. The problem of estimating future demand in a technologically changing world should best be approached from a cybernetic viewpoint; here however we briefly consider the problem of determining the optimal capacity of a system, where demand and supply functions (and their p.d.f.'s) are known. Work on this problem has been done by various authors, e.g. Hurst (1956), Koopmans (1957), Arrow, Karlin and Scarf (1958), Prabhu (1965), Ghosal (1966), Ghosal, Aylawadih and Chadda (1967), Bose (1967), etc.

4.2 Hurst's Work on Reservoir Capacity

Hurst's problem was to determine the optimal capcaity of k for a dam governed by the model (1.2); he calculated $U_i = \sum_{j=1}^{i} u_j$, where $u_j = X_j - Y_j$, in most of the dam models Y_j being assumed to be m. He calculated the range (T being the time horizon under review)

$$R_T = \max_{1\le i\le T} U_i - \min_{1\le i\le T} U_i \tag{4.1}$$

and derived approximately

$$E(R_T) = \sigma\sqrt{T\pi/2} \tag{4.2}$$

where $\sigma^2 = \text{var}(u_t)$. If $Y_t = m$, $\sigma^2 = \text{var}(X_t)$; if $X_t = m$ (constant), $\sigma^2 = \text{var}(Y_t)$. His thesis was that the optimal capacity of a reservoir should be $E(R_T)$. This procedure often leads to creation of an unusually large capacity if we take T to be large, because $E(R_T)$ increases with the increase in T. Secondly it does not take into consideration any financial factor (cost or benefit); this may however be justified in taking a decision for the community which is expected to derive considerable social return in future.

4.3 Method based on Risk of Emptiness

We may develop a simple method for deriving the approximate cap city k, if we know the policy which is likely to be followed in operating the system (dam or an inventory system) and the p.d.f.'s of X_t and Y_t (one of these maybbe deterministic). If we define V(y,k) to be the probability that dam becomes empty, when the initial level is y and the dam capacity is k, then the lower the value of V(y,k) the better it is for the community. For each policy of operation we can plot a set of values for V(y,k) for various values of y and k. We may subjectively derive k and the risk level (y_r) by adopting the criterion, say $V(y,k) < 0.80$. The method of deriving $V(y,k)$ and its relation to the limiting d.f. of Z_t have been discussed in Chapter 2. The practical utility of this method is that, where we cannot derive $V(y,k)$ analytically, we can study this function by numerical methods: the plotting of $V(y,k)$ against y (for same k) for various policies gives an insight into the optimal policies to be adopted at different situations. Again the plotting of a scatter (y,k) for a particular value of $V(y,k)$ (say = 0.80) gives an insight into the optimal capacity (k) to be built in the light of available resources (see Ghosal (1969), Chapter 3).

4.4 An Algorithm for determining optimal k

Let us assume that a storage system $\langle Z_t, X_t, Y_t, k\rangle$ or $\langle Z_t, u_t, k\rangle$ reach a stationary state: in other words u_t follows a d.f. $G(x)$, $0 \leq x < \infty$, and Z_t attains a limiting d.f. $F(y)$, $0 \leq y \leq k$. We assume t to be discrete, but the continuous time-model can be easily developed. We have

$$Z_{t+1} = \min[Z_t + X_t - Y_t, k]^+;$$

in a dam model we may assume Y_t to be deterministic (= m), X_t an r.v.; in an inventory situation X_t may be deterministic and Y_t an r.v. Let us introduce the following cost factors:

$c(k)$ = the capital cost of erecting a storage capacity k;

$h(k)$ = the cost of maintenance of a system of capacity k per unit time;

β the revenue from sotring a unit over unit time (week, month, year, etc),

γ = penalty cost for not providing storage space for a unit item over unit time.

We may have $c(k)$, $h(k)$ an linear or non-linear function of k; here for the sake of simplicity we assume a linear model. We neglect γ, the penalty cost, for the time being.

We determine the optimal k such that the estimate of discounted value of net earnings(= gross earning - cost) over a horizon of time (0,T) is maximum. We may evaluate cost and benefit in terms of social returns. The present value of cost, both maintenance and capital costs combined, over the period (0,T) is given by

$$C(k,T) = c(k) + h(k)a_{\overline{T|}} \tag{4.3}$$

where $a_{\overline{T|}} = 1 - v^t/i$, i being the rate of interest per unit time and $v = 1/(1+i)$. The present value of the estimated revenue over the period (0,T) is given by

$$R(k,T) \sim \sum_{t=1}^{T} E(Z_t)\, v^t$$

$$= \beta E(Z_t) a_{\overline{T|}} \tag{4.4}$$

if $E(Z_t)$ is the same for all t. Thus, the present value of net benefit B(k,T) is given by

$$B(k,T) = \beta E(Z;k)\, a_{\overline{T|}} - c(k) - h(k)a_{\overline{T|}} \ , \tag{4.5}$$

where E(Z;k) denotes the expectation of Z if a capacity k is avaialable. We know

$$E(Z;k) = \int_{0-}^{k+} y \, dF(y;k) = \int_0^k \overline{F}(y,k)dy \tag{4.6}$$

where $F(y;k) = pr(Z_t \leq y, \text{ given } k)$, $\overline{F}(y,k) = 1 - F(y,k)$. Thus we have

$$E(Z;k + \Delta k) = \int_0^{k+\Delta k} \overline{F}(y;k + \Delta k)dy$$

so that

$$E(Z;k+\Delta k) - E(Z;k) = \int_0^{k+\Delta k} \overline{F}(y;k+\Delta k)dy - \int_0^k \overline{F}(y;k)dy$$

$$= \int_0^k \{\overline{F}(y;k+\Delta k) - \overline{F}(y;k)\}dy + \int_k^{k+\Delta k} \overline{F}y;k+\Delta k)dy;$$

the left-hand term on the R.H.S. is generally very small, so that for all practical purposes, we may assume approximately (particularly because there is a point probability at k)

$$E(Z;k+\Delta k) - E(Z;k) \to \overline{F}(k;k+\Delta k)\Delta k \to F(k)\Delta k.$$
$$= f(k;k)\Delta k$$

where $f(k;k)$ is the concentration at point k when the capacity is k. Thus approximately we may assume

$$dE(Z;k)/dk \to f(k;k). \qquad (4.7)$$

Now differentiating $B(k;T)$ in (4.6) w.r.t. k, and equating to zero, we get

$$0 = dB(k,T)/dT$$
$$= \beta\{dE(z;k)/dk\}a_T - c!(k) - h'(k)\ a_{\overline{T}|}$$
$$\to \beta f(k,k)a_T - c'(k) - h'(k)\ a_{\overline{T}|}\ , \qquad (4.8)$$

so that

$$f(k;k) \to \frac{c'(k) + h'(k)\ a_T}{\beta a_T}\ . \qquad (4.9)$$

We have to determine k such that (4.9) is satisfied. We have to determine by numerical methods. If $c(k)$ is linear $c(k) = ck$, $c'(k) = c$; but if we know the exact functional form of $c(k)$ and $h(k)$ then also we can determine k from the above.

The algorithm (4.9) is therefore based on the determination of $f(k;k)$, which we can do analytically for particular d.f.'s of X_t (or Y_t) (see Chapter 1). If we find that the approximation (4.7) is not valid, we may obtain $dE(\cdot)/dk$ directly, particularly if X_t (or Y_t) follows a discrete d.f. For example if Y_t follows a negative binomial d.f. and $X_t = m$ (Ghosal, 1966):

$$g_j = pr(Y_t = i) = \binom{p+j-1}{j} a^p b^j, \quad a = 1 - b,\ p > 0,$$
$$j = 0,1,\ldots \qquad (4.10)$$

then we can determine $\overline{F}_i$ directly by solving a set of linear equations (eq. (12) in Ghosal, 1966) and then calculate E(Z;k) directly numerically, The algorithm (4.9) was given by Ghosal (1966). Models involving penalty cost are being studied by Aylawadi and Ghosal (unpublished).

4.5 (s,S) Policy

The (s,S) inventory policy has been studied in Chapter 1 (equation (1.104), p.59); we have

$$Z_{t+1} = Z_t - Y_t \quad \text{if} \quad s < Z_t \leq S,$$
$$= S - Y_t \quad \text{if} \quad Z_t \leq s.$$

The problem may be to determine s and S optimally with respect to some cost criterion, the d.f. of Y_t being known. This problem has been studied by Arrow, Karlin and Scarf (1958) and Prabhu (1965). If we denote the ordering cost of quantity x ans cs + K, for x > 0, K being constant, penalty cost per unit shortage as p, shortage cost per unit as h, then the cost function associated with the (s,S) policy for the level Z_t is (G(x) is the d.f. of Y_t)

$$L(Z_t) = \begin{cases} K + c(S-Z_t) + p\int_S^\infty (x-S)\,dG(x) \\ \qquad + h\int_0^S (S-x)\,dG(x), \quad Z_t \leq s, \\ \\ p\int_{Z_t}^\infty (x-Z_t)\,dG(x) + h\int_0^{Z_t} (Z_t-x)\,dG(x) \\ \qquad s < Z_t \leq S \end{cases} \qquad (4.11)$$

The expected cost when the process has settled down to an equilibrium is given by

$$L(s,S) = \int_{-\infty}^{S} L(x) \lim_{t\to\infty} dF_t(x),$$

$$= \int_{0}^{S} L(x)\, dF(x) \qquad (4.12)$$

The optimal policy is the one ($s = s_o$, $S = S_o$) which make $L(s,S)$ minimum. Here also we have to resort to extensive computer calculations.

We can determine (s,S) analytically for simple d.f.'s of Y_t, viz. exponential, geometric, etc. Bose (1967) has applied the method given in Section 4.4 for this problem.

4.6 Concluding Remarks.

There is a host of optimization problems with respect to storage (or queueing) systems, all of which could not be discussed here. A very significant problem in this field was the one discussed elegantly by Koopmans (1957): his problem was to devise a method for optimizing a policy of managing a hydro-electric system such that the cost of generation of thermal power (= total power requirement - hydro-eleectric power generated) was minimum over a time horizon $(0,T)$ - given the input into the dam and its output. A stochastic analogue of Koopmans' model is being worked out by the author, and the results will be reported subsequently.

BIBLIOGRAPHY

1. Arrow, K.J., Karlin, S. and Scarf, H. (1958). Studies in the Mathematical Theory of Inventory and Production. Stanford Univ. Press.

2. Bailey, Norman T.J. (1954). On queueing process with bulk service. J. Roy. Stat. Soc. B16, 80-87.

3. G. Baxter and M. Donsker (1957). On the supremum functional for processes with stationary independent increments. Trans. Amer. Soc. 85 113-123.

4. Bose. K. (1967). A note on determination of the optimum capacity of a storage system. Opsearch 4, 175-178.

5. Brown, Bernice (1965). Some Tables of the Negative Binomial Distribution and their use. Rand Corpn.

6. Cox D.R. and Smith, W.L. (1961). Queues Methuen, London.

7. Finch, P.D. (1958). The effect of the size of the waiting room on a simple queue. J. Roy. Stat. Soc. B20 182-186

8. Finch, P.D. (1959). On the distribution of queue size in queueing problems. Acta Math. Acad. Sci. Hung. 10, 327-336.

9. Finch, P.D. (1960). Deterministic customer impatience in the queueing system GI/M/1. Biometrika 47, 45-52.

10. Finch, P.D. (1961). On the busy period in the queueing system GI/G/1. J. Aust. Math. Soc. 2, 217-228.

11. Gani, J. and Prabhu, N.U. (1957). Stationary distributions of the negative exponential type of input for the infinite dam. J. Roy. Statist. Soc. 19B 342-351

12. Ghosal, A. (1959). On the continuous analogue of Holdaway's problem for the finite dam. Austr. J. Appl. Sci. 10, 365-370.

13. Ghosal A. (1960). Emptiness in the finite dam. Ann. Math. Statist. 31, 803-808.

14. Ghosal, A. (1962a). Contributions to the Theory of Dams. D. Phil. Thesis. Calcutta Univ. (1962, Unpublished)

15. Ghosal, A. (1962b). Finite Dam with negative binomial input. Austr. J. Appl. Sci. 13 71-74.

16. Ghosal, A. (1962c). Queues in series. J. Roy. Statist. Soc. 24B, 359-363.

17. Ghosal, A. (1963a). Stochastic treatment of inventory problems. Colloq. on Appl. of Math. to Economics Academy of Sci. Hungary Budapest. (Proc. published 1965).

18. Ghosal, A. (1963b). Queues with finite waiting time. Operations Res. 11 919-921.

19. Ghosal, A. (1964). Some results in the theory of inventory. Biometrika 51 487-490.

20. Ghosal, A. (1966). An Inventory Problem: Determination of the Optimum Capacity of a Storage System. Opsearch 3, 1-13.

21. Ghosal, A., Aylawadi, D.R., Chadda, S.S. (1967). "On the determination of optimum capacity of a storage system" (Paper in) Advancing Frontiers in Operational Research (Proc. Internat. Symp. on O.R. New Delhi) Hindusthan Pub. Corpn. Delhi.

22. Ghosal, A. (1968). Some remarks about duality relations in queues Acta Math. Acd. Sci. Hung. 19, 163-170.

23. Ghosal, A. (1969). Elements of Operational Research Hindusthan Publishing Corpn, Delhi.

24. Hurst, H.E. (1956). Methods of using long-term storage in reservoirs Proc. Inst. Civ. Engrs. Paper 6059. 5 Pt I, 519-590.

25. Jackson, R.R.P. (1956). Queueing processes with phase-type service. J. Roy. Stat. Soc. B18, 129-132.

26. Keilson, J. (1966). The ergodic queue length distribution for queueing with finite capacity. J. Roy. Statist. Soc. B28, 190-201

27. Kendall, David G. (1951). Some problems in the theory of queues. J. Roy. Statist. Soc. B13, 151-185.

28. Kendall, David G. (1953). Stochastic processes occurring in the theory of queues and their analysis by the method of the imbedded Markov chain. Ann. Math. Statist. 24, 338-354.

29. Kendall, David G. (1957). Some results in the theory of dams. J. Roy. Statist. Soc. B19, 207-212.

30. Kingman, J.F.C. (1966). On the algebra of queues. J. Appl. Prob. 3, 285-326.

31. Klir, J. and Valach, M. (1965). Cybernetic Modelling. (English Translation Edn 1967) Illiffe Books, London.

32. Koppmans, Tjalling C. (1958). Water Storage Policy in a Simplified Hydro-electric System. Cowles Commn. Paper No. 115, Yales Univ.

33. Le Gall, P. (1962). Les Systemes avec ou sans attente, et les processus stochastiques Tome I, Dunod.

34. Lindley, D.V. (1952). The theory of queues with a single server. Proc. Camb. Phil. Soc. 48, 277-289.

35. Moran, P.A.P. (1954). A probability theory of dams and Storage systems. Aust. J. Appl. Sci. 5, 116-124.

36. Moran, P.A.P. (1955). A probability theory of dams and storage systems: modification of release rules. Aust. J. App. Sci. 6, 177-130.

37. Moran, P.A.P. (1958). The Theory of Storage. Methuens, London.

38. Morse, P.M. (1958). Queues, Inventories and Maintenance. John Wiley: New York.

39. Nelson, Rosser T. (1966). Queueing Network Experiments with varying arrival and service processes. Naval Res. Logistics Qtly 13, 321-347.

40. Pollaczek, F. (1957). Problémes stochastiques poses par le phénomène de formation d'une queue d'attente à un guichet et par des phenomenes apparentés (Mémorial des Sciences Mathématiques, fasc. 136) Gauthier-Villars, Paris.

41. Prabhu, N.U. (1958a). On the integral equation for the finite dam. Quart. J. Math. (Oxford) 9, 183-188.

42. Prabhu, N.U. (1958b). Some exact results for the finite dam. Ann. Math. Stat 29, 1234-1243.

43. Prabhu, N.U. (1960). Some results for the queue with Poisson arrivals. J. Roy. Stat. Soc B22, 104-107.

44. Prabhu, N.U. (1962). Elementary methods for some waiting time problems. Operations Res. 10, 559-566.

45. Prabhu, N.U. (1965). Queues and Inventory. John Wiley New York.

46. Rosenblatt, M. (1954). An inventory problem. Econometrica 22, 244-247.

47. Saaty, T.L. (1964). Stochastic network flows: advances in network of queues. Proc. Symp. on Congestion Theory Univ. of North Carolina 1966, 86-107.

48. Smith, Walter L. (1953). On the distribution of queueing times. Proc. Camb. Phil. Soc. 49, 449-461.

49. Sparre-Andersen, E. (1954). On the fluctuations of sums of random variables II. Math. Scand. 2, 195-223.

50. Spitzer, Frank.(1956). A Combinatorial Lemma and its application to probability theory. Trans. Amer. Math. Soc. 82, 449-461.

51. Spitzer, Frank.(1957). The Wiener-Hopf Equation whose kernel is a probability density. Duke Math. J. 24, 327-343.

52. Tocher, K.D. (1963). The Art of Simulation. The English Univ. Press Lond.

53. Wendel, J.G. (1958). Spitzer's formula: a short proof. Proc. Amer. Math. Soc. 9, 905-908.

Offsetdruck: Julius Beltz, Weinheim/Bergstr.

Lecture Notes in Operations Research and Mathematical Systems

Vol. 1: H. Bühlmann, H. Loeffel, E. Nievergelt, Einführung in die Theorie und Praxis der Entscheidung bei Unsicherheit. 2. Auflage, IV, 125 Seiten 4°. 1969. DM 12,– / US $ 3.30

Vol. 2: U. N. Bhat, A Study of the Queueing Systems M/G/1 and GI/M/1. VIII, 78 pages. 4°. 1968. DM 8,80 / US $ 2.50

Vol. 3: A. Strauss, An Introduction to Optimal Control Theory. VI, 153 pages. 4°. 1968. DM 14,– / US $ 3.90

Vol. 4: Einführung in die Methode Branch and Bound. Herausgegeben von F. Weinberg. VIII, 159 Seiten. 4°. 1968. DM 14,– / US $ 3.90

Vol. 5: L. Hyvärinen, Information Theory for Systems Engineers. VIII, 295 pages. 4°. 1968. DM 15,20 / US $ 4.20

Vol. 6: H. P. Künzi, O. Müller, E. Nievergelt, Einführungskursus in die dynamische Programmierung. IV, 103 Seiten. 4°. 1968. DM 9,– / US $ 2.50

Vol. 7: W. Popp, Einführung in die Theorie der Lagerhaltung. VI, 173 Seiten. 4°. 1968. DM 14,80 / US $ 4.10

Vol. 8: J. Teghem, J. Loris-Teghem, J. P. Lambotte, Modèles d'Attente M/G/1 et GI/M/1 à Arrivées et Services en Groupes. IV, 53 pages. 4°. 1969. DM 6,– / US $ 1.70

Vol. 9: E. Schultze, Einführung in die mathematischen Grundlagen der Informationstheorie. VI, 116 Seiten. 4°. 1969. DM 10,– / US $ 2.80

Vol. 10: D. Hochstädter, Stochastische Lagerhaltungsmodelle. VI, 269 Seiten. 4°. 1969. DM 18,– / US $ 5.00

Vol. 11/12: Mathematical Systems Theory and Economics. Edited by H. W. Kuhn and G. P. Szegö. VIII, IV, 486 pages. 4°. 1969. DM 34,– / US $ 9.40

Vol. 13: Heuristische Planungsmethoden. Herausgegeben von F. Weinberg und C. A. Zehnder. II, 93 Seiten. 4°. 1969. DM 8,– / US $ 2.20

Vol. 14: Computing Methods in Optimization Problems. Edited by A. V. Balakrishnan. V, 191 pages. 4°. 1969. DM 14,– / US $ 3.90

Vol. 15: Economic Models, Estimation and Risk Programming: Essays in Honor of Gerhard Tintner. Edited by K. A. Fox, G. V. L. Narasimham and J. K. Sengupta. VIII, 461 pages. 4°. 1969. DM 24,– / US $ 6.60

Vol. 16: H. P. Künzi und W. Oettli, Nichtlineare Optimierung: Neuere Verfahren, Bibliographie. IV, 180 Seiten. 4°. 1969. DM 12,– / US $ 3.30

Vol. 17: H. Bauer und K. Neumann, Berechnung optimaler Steuerungen, Maximumprinzip und dynamische Optimierung. VIII, 188 Seiten. 4°. 1969. DM 14,– / US $ 3.90

Vol. 18: M. Wolff, Optimale Instandhaltungspolitiken in einfachen Systemen. V, 143 Seiten. 4°. 1970. DM 12,– / US $ 3.30

Vol. 19: L. Hyvärinen, Mathematical Modeling for Industrial Processes. VI, 122 pages. 4°. 1970. DM 10,– / US $ 2.80

Vol. 20: G. Uebe, Optimale Fahrpläne. IX, 161 Seiten. 4°. 1970. DM 12,– / US $ 3.30

Vol. 21: Th. Liebling, Graphentheorie in Planungs- und Tourenproblemen am Beispiel des städtischen Straßendienstes. IX, 118 Seiten. 4°. 1970. DM 12,– / US $ 3.30

Vol. 22: W. Eichhorn, Theorie der homogenen Produktionsfunktion. VIII, 119 Seiten. 4°. 1970. DM 12,– / US $ 3.30

Vol. 23: A. Ghosal, Some Aspects of Queueing and Storage Systems. IV, 93 pages. 4°. 1970. DM 10,– / US $ 2.80

Beschaffenheit der Manuskripte

Die Manuskripte werden photomechanisch vervielfältigt; sie müssen daher in sauberer Schreibmaschinenschrift geschrieben sein. Handschriftliche Formeln bitte nur mit schwarzer Tusche eintragen. Notwendige Korrekturen sind bei dem bereits geschriebenen Text entweder durch Überkleben des alten Textes vorzunehmen oder aber müssen die zu korrigierenden Stellen mit weißem Korrekturlack abgedeckt werden. Falls das Manuskript oder Teile desselben neu geschrieben werden müssen, ist der Verlag bereit, dem Autor bei Erscheinen seines Bandes einen angemessenen Betrag zu zahlen. Die Autoren erhalten 75 Freiexemplare.

Zur Erreichung eines möglichst optimalen Reproduktionsergebnisses ist es erwünscht, daß bei der vorgesehenen Verkleinerung der Manuskripte der Text auf einer Seite in der Breite möglichst 18 cm und in der Höhe 26,5 cm nicht überschreitet. Entsprechende Satzspiegelvordrucke werden vom Verlag gern auf Anforderung zur Verfügung gestellt.

Manuskripte, in englischer, deutscher oder französischer Sprache abgefaßt, nimmt Prof. Dr. M. Beckmann, Department of Economics, Brown University, Providence, Rhode Island 02912/USA oder Prof. Dr. H. P. Künzi, Institut für Operations Research und elektronische Datenverarbeitung der Universität Zürich, Sumatrastraße 30, 8006 Zürich entgegen.

Cette série a pour but de donner des informations rapides, de niveau élevé, sur des développements récents en économétrie mathématique et en recherche opérationnelle, aussi bien dans la recherche que dans l'enseignement supérieur. On prévoit de publier

1. des versions préliminaires de travaux originaux et de monographies
2. des cours spéciaux portant sur un domaine nouveau ou sur des aspects nouveaux de domaines classiques
3. des rapports de séminaires
4. des conférences faites à des congrès ou à des colloquiums

En outre il est prévu de publier dans cette série, si la demande le justifie, des rapports de séminaires et des cours multicopiés ailleurs mais déjà épuisés.

Dans l'intérêt d'une diffusion rapide, les contributions auront souvent un caractère provisoire; le cas échéant, les démonstrations ne seront données que dans les grandes lignes. Les travaux présentés pourront également paraître ailleurs. Une réserve suffisante d'exemplaires sera toujours disponible. En permettant aux personnes intéressées d'être informées plus rapidement, les éditeurs Springer espèrent, par cette série de »prépublications«, rendre d'appréciables services aux instituts de mathématiques. Les annonces dans les revues spécialisées, les inscriptions aux catalogues et les copyrights rendront plus facile aux bibliothèques la tâche de réunir une documentation complète.

Présentation des manuscrits

Les manuscrits, étant reproduits par procédé photomécanique, doivent être soigneusement dactylographiés. Il est recommandé d'écrire à l'encre de Chine noire les formules non dactylographiées. Les corrections nécessaires doivent être effectuées soit par collage du nouveau texte sur l'ancien soit en recouvrant les endroits à corriger par du verni correcteur blanc.

S'il s'avère nécessaire d'écrire de nouveau le manuscrit, soit complètement, soit en partie, la maison d'édition se déclare prête à verser à l'auteur, lors de la parution du volume, le montant des frais correspondants. Les auteurs recoivent 75 exemplaires gratuits.

Pour obtenir une reproduction optimale il est désirable que le texte dactylographié sur une page ne dépasse pas 26,5 cm en hauteur et 18 cm en largeur. Sur demande la maison d'édition met à la disposition des auteurs du papier spécialement préparé.

Les manuscrits en anglais, allemand ou francais peuvent être adressés au Prof. Dr. M. Beckmann, Department of Economics, Brown University, Providence, Rhode Island 02912/USA ou au Prof. Dr. H.P. Künzi, Institut für Operations Research und elektronische Datenverarbeitung der Universität Zürich, Sumatrastraße 30, 8006 Zürich.

GPSR Compliance

The European Union's (EU) General Product Safety Regulation (GPSR) is a set of rules that requires consumer products to be safe and our obligations to ensure this.

If you have any concerns about our products, you can contact us on ProductSafety@springernature.com

In case Publisher is established outside the EU, the EU authorized representative is:

Springer Nature Customer Service Center GmbH
Europaplatz 3
69115 Heidelberg, Germany

Batch number: 10845480

Printed by Printforce, the Netherlands